MATHSWISE

Book Three

Ray Allan
Henry Compton School, Fulham.

Martin Williams
Henry Compton School, Fulham.

Oxford University Press

Oxford University Press, Walton Street, Oxford OX2 6DP

Oxford New York Toronto
Delhi Bombay Calcutta Madras Karachi
Kuala Lumpur Singapore Hong Kong Tokyo
Nairobi Dar es Salaam Cape Town
Melbourne Auckland Madrid

and associated companies in
Berlin Ibadan

Oxford is a trade mark of Oxford University Press

ISBN 0 19 834769 3

© Oxford University Press 1985

First published 1985
Reprinted 1987, 1988, 1990, 1991, 1992, 1993

The authors and the publisher are grateful to Guinness
Superlatives Ltd for information obtained on various world
records which are inserted in this book. Readers are invited
to consult the current edition of *The Guiness Book of Records*
for the latest information on these and other records.

Cover illustration by Terry Pastor
Text illustrations by Jon Riley

Artwork and Typesetting by BBG Ltd., Bristol
Printed in Italy by G. Canale & C. S.p.A. - Turin

Contents

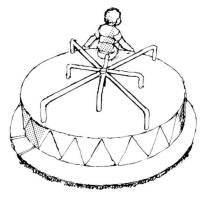

Jenny takes her little brother Tommy to the park. She puts him on the roundabout and slowly turns it.

Below you can see Tommy turning on the roundabout.

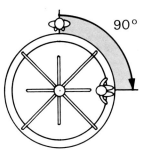

90°

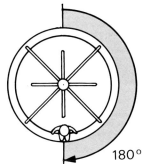

180°

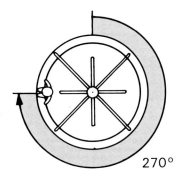

270°

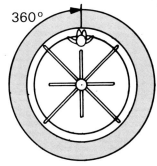

360°

Here Tommy has turned through one right angle or 90°

Here Tommy has turned through two right angles or 180°

Here Tommy has turned through three right angles or 270°

Here Tommy has made one turn. This is four right angles or 360°

Exercise 1

There are 360° or four right angles in one turn.

How many more degrees of turn are needed to complete one turn or 360°?

1.

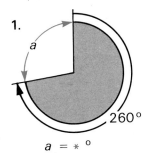

a

260°

$a = *°$

2.

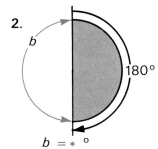

b

180°

$b = *°$

3.

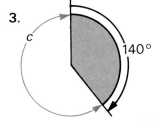

c

140°

$c = *°$

4.

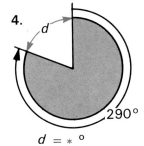

d

290°

$d = *°$

5.

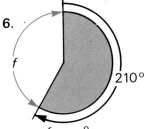

70°

e

$e = *°$

6.

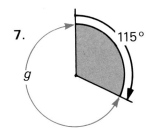

f

210°

$f = *°$

7.

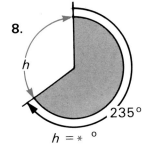

115°

g

$g = *°$

8.

235°

h

$h = *°$

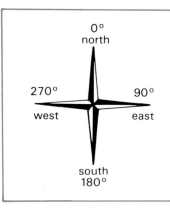

This is a compass rose. It shows the positions north, east, west and south.

The angle between each of the four compass points is 90° or one right angle.

We always count angles on a compass in a clockwise direction starting from north (0°).
So east is 90° from north, south is 180° from north and so on.

Exercise 2

Copy and complete the sentences below.

1. There are ____° between north and east.

2. There are ____° between north and south.

3. There are ____° between north and west.

4. One full turn is ____° or ____ right angles.

5. From east to south is ____° (remember to turn clockwise).

6. From east to north is ____°.

7. If you turn from south to north, you turn through ____°.

8. There are ____° from west to south.

9. One half of a full turn is ____° or ____ right angles.

Exercise 3

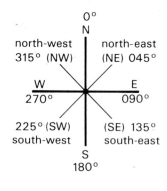

Use the compass rose below to complete these sentences.

1. There are ____° between north and north-east.

2. There are ____° between north and south-east.

3. From north to north-west is ____°

4. From north-east to south-east is ____°

5. There are ____° between south-east and north-west.

6. If you turn from west to north-east you have turned through ____°

7. Complete the sentences below using either the word 'more' or the word 'less'.

   a. A turn from north to north-east is ____ than one right angle.

   b. The angle between south-west and east is ____ than two right angles.

   c. From north to north-west is ____ than four right angles.

   d. A full turn is ____ than three right angles.

Exercise 4

Mary is standing in the middle of this roundabout. What will she see if she looks:—

1. east?
2. north?
3. south-west?
4. north-west?

What direction is Mary facing if she can see:—

5. Cliff Road?
6. Ice Lane?
7. The Station?
8. Castle Street?

9. A man standing outside the Station sees Mary. In which direction is he looking?

10. A woman standing outside the Museum sees Mary. In which direction is she looking?

11. Wallace Way is a one-way-street. In which direction does the traffic travel?

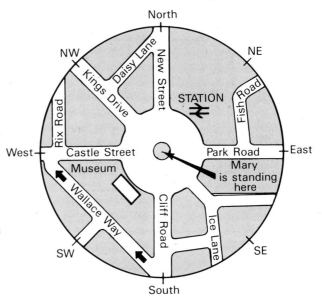

Exercise 5

Here is the radar screen on H.M.S. Tiger.
It shows other ships, objects and land.
H.M.S. Tiger is heading on a bearing of 090°.

Use a ruler to help you work out the bearings of the following points.

OBJECT	BEARING
H.M.S. Argos	* * *
Felix Island	* * *
rocks	* * *
H.M.S. Lark	* * *
H.M.S. Hero	* * *
lightship	* * *
*	225°
*	300°
*	110°
*	165°

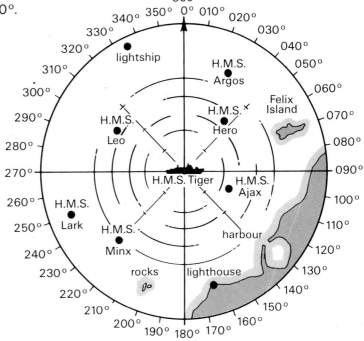

1. The harbour is on a bearing of _____°

2. How many ships are sailing to the north of H.M.S. Tiger?

3. What will happen if H.M.S. Tiger continues on a course of 090°?

4. Which ship is roughly north-west of H.M.S. Tiger?

Exercise 6

1. Trace or draw these angles accurately onto card and cut them out.

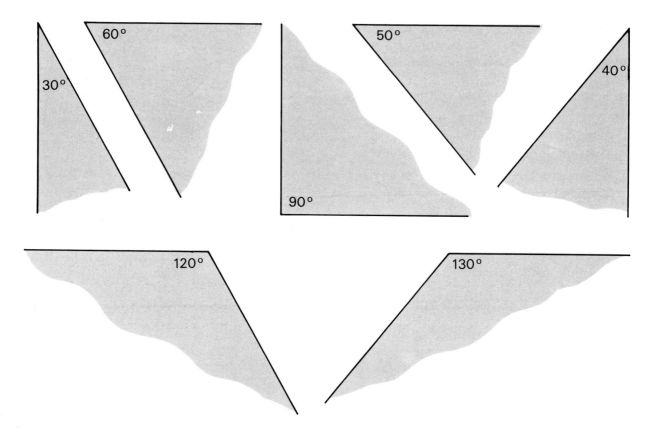

2. Put the 50°, 40° and 90° angle cards together so they form a straight line. Check that the cards form a straight line by using the edge of a ruler.

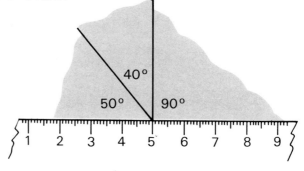

Exercise 7

Use the angle cards to answer these questions
a. Which groups of angles below form straight lines?
b. Add up the angles in the groups which form straight lines.

1. Group A: 40°, 90°, 50°

2. Group B: 130°, 40°

3. Group C: 60°, 120°

4. Group D: 120°, 130°

5. Group E: 120°, 60°

6. Group F: 50°, 130°

7. Group G: 60°, 30°, 90°

8. Group H: 50°, 60°, 30°, 40°

Angles on a straight line

Exercise 8

Use a protractor to find the angle of the
ladder in each drawing.
The first one is done for you.

1.

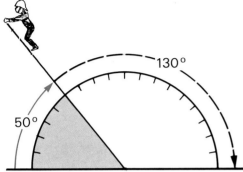

The ladder has been raised by 50°. It would
have to turn through another 130° to complete
half a turn.

2.

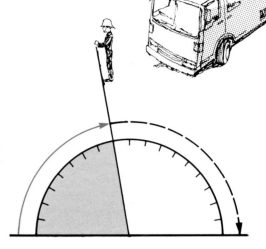

The ladder has been raised by ___°. It would
have to turn through another ___° to complete
half a turn.

3.

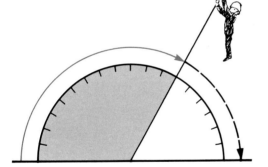

The ladder has been raised by ___°. It would
have to turn through another ___° to complete
half a turn.

4.

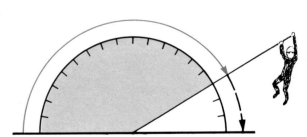

The ladder has been raised by ___°. It would
have to turn through another ___° to complete
half a turn.

Exercise 9

Work out the missing angle in each drawing. Remember that the angles
must add up to 180°.

1.

140° a

2.

60° b

3.

100° c

4.

20° d

5.

50°
50° e

6.

f
60°   70°

7.

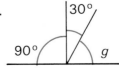

30°
90° g

8.

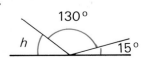

130°
h 15°

Parallel lines are the same distance away from each other at every point.

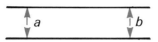 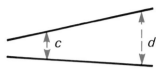

These lines are parallel. The width at *a* is the same as at *b*.

These lines are not parallel. The distance at *c* is less than at *d*.

The sides of this ladder are the same distance apart at every point. The sides are parallel.

The sides of this ladder are not the same distance apart at every point. (The sides get closer at the top.) The sides are not parallel.

Exercise 1

Which of these pairs of lines are parallel? Use a ruler to find out if the lines are the same distance apart at every point.

1.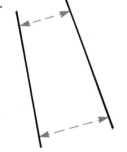

2.

3.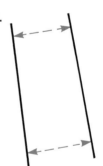

4.

5.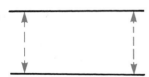

6.

7.

8.

Exercise 2

Which line, a, b, c or d is parallel to the line drawn below in blue?
Use a ruler to help you.

1.   a
      b
      c
      d

r ——————————— r

2.   a
      b
      c
      d

s
      s

3.   a
      b
      c
      d

t
t

4.   a
      b
      c
      d

u
      u

5.   a
      b
      c
      d

w
      w

6.   a
      b
      c
      d

x ——————————— x

Exercise 3

1. Draw a 5 cm line across your page; 2 cm below it draw another 5 cm line, parallel to your first line.

2. Draw a 4 cm line across your page; 1 cm below it draw another 4 cm line, parallel to your first line.

3. Draw a 6 cm line across your page; 3 cm below it draw another 6 cm line, parallel to your first line.

4. Draw a 5 cm line across your page; 2·5 cm below it draw another 5 cm line, parallel to your first line.

5. Draw a 4·5 cm line across your page; 1·5 cm below it draw another 4·5 line, parallel to your first line.

6. The pair of heavy lines in one of these drawings is not parallel. Which drawing is it?

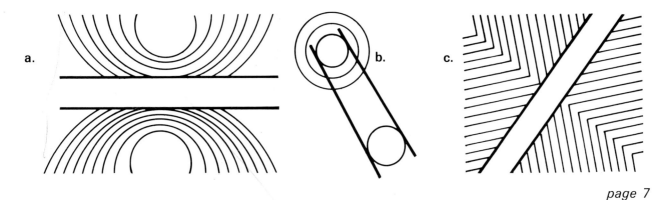

a.   b.   c.

page 7

Angles and parallel lines

Exercise 4   Use a protractor to help you to answer these questions.

1. **a.** Angle A = ____ °
   b. Angle B = ____ °
   c. Are the two angles the same size?
   d. Are the lines a and b parallel?

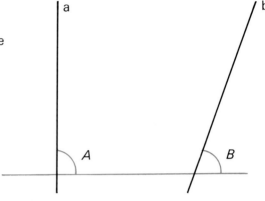

2. **a.** Angle C = ____ °
   b. Angle D = ____ °
   c. Are the two angles the same size?
   d. Are the lines c and d parallel?

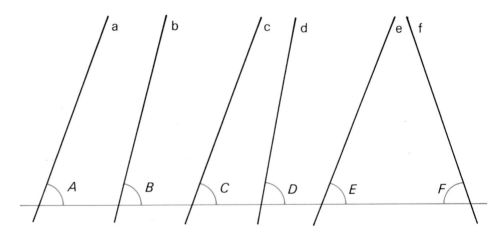

3. **a.** Angle A = * °   **b.** Angle B = * °   **c.** Angle C = * °
   d. Angle D = * °   **e.** Angle E = * °   **f.** Angle F = * °
   g. Which angles are the same size as angle A?
   h. Which lines are parallel to line a?
   i. Is line b parallel to line d?

Exercise 5   By comparing the angles in each diagram below, say which drawings show a pair of parallel lines.

1.

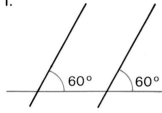

60°   60°

2.

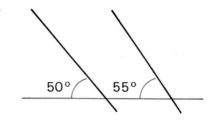

50°   55°

3.

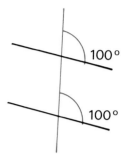

100°

100°

4.

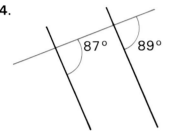

87°   89°

5.

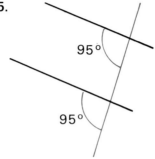

95°

95°

6.

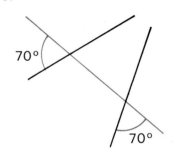

70°

70°

7. What is the distance marked *d*?

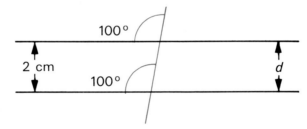

100°

2 cm

100°

d

8. What is the size of the angle marked *A*?

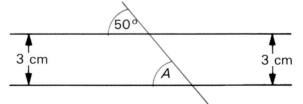

50°

3 cm

3 cm

A

9. John is writing the size of an angle onto his diagram.
   a. Is John writing in the correct size of angle?

   b. If your answer is no, say why John's angle is wrong.

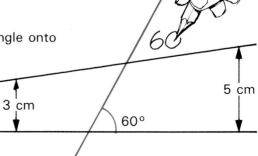

60

3 cm

60°

5 cm

Section 3　Sets

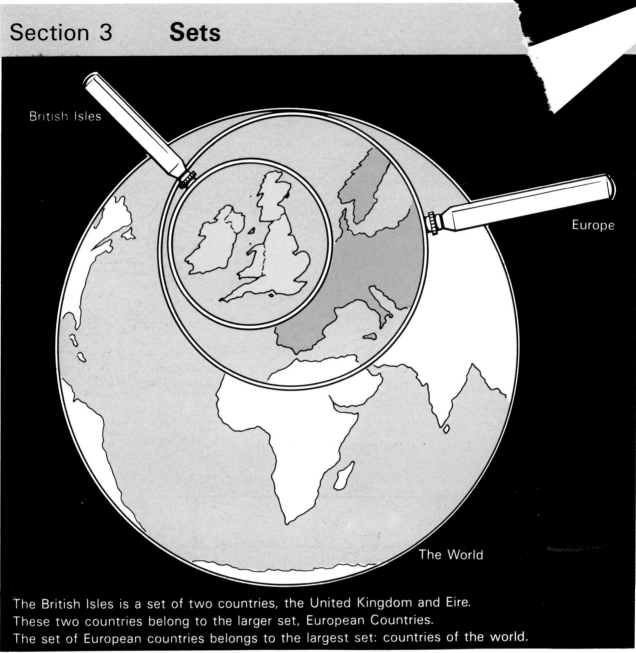

British Isles

Europe

The World

The British Isles is a set of two countries, the United Kingdom and Eire.
These two countries belong to the larger set, European Countries.
The set of European countries belongs to the largest set: countries of the world.

On a diagram we show this information thus:

W is the countries of the world.
E is the countries of Europe.
B is the countries of the British Isles.

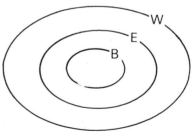

Exercise 1　　　　Draw set rings like the ones above. Put these countries into their correct
positions on the diagram.

1. India　　2. France　　3. Canada　　4. The United Kingdom
5. Italy　　6. New Zealand　　7. Jamaica　　8. Egypt

Here is a group of shapes.

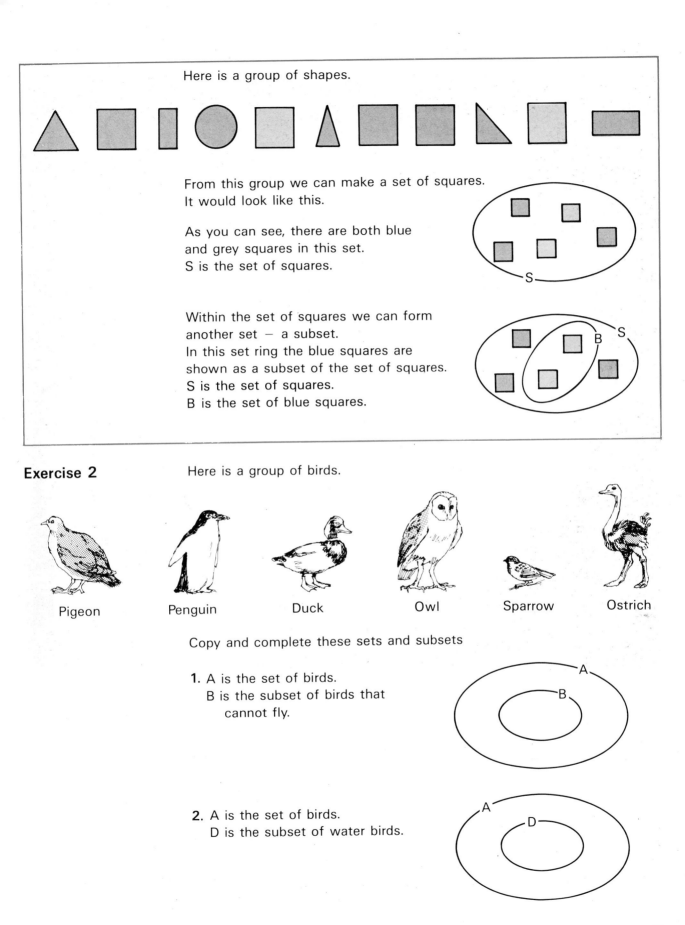

From this group we can make a set of squares.
It would look like this.

As you can see, there are both blue
and grey squares in this set.
S is the set of squares.

Within the set of squares we can form
another set — a subset.
In this set ring the blue squares are
shown as a subset of the set of squares.
S is the set of squares.
B is the set of blue squares.

Exercise 2

Here is a group of birds.

Pigeon   Penguin   Duck   Owl   Sparrow   Ostrich

Copy and complete these sets and subsets

1. A is the set of birds.
   B is the subset of birds that
      cannot fly.

2. A is the set of birds.
   D is the subset of water birds.

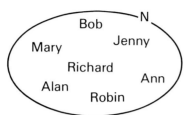
N is the set of names.

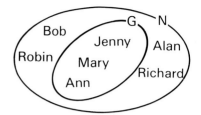
G is the subset of girls' names

Exercise 3

Copy these set rings and their members.

1. M is the set of units of measurement

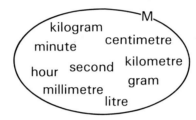

Put a ring around units that measure time.
Call this subset T.

2. C is the set of colours

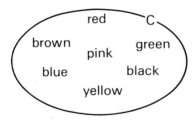

Put a ring around colours that begin with the letter b
Call this subset B.

Exercise 4

Copy this set ring and its members three times.

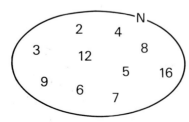

1. Within this set form a subset of
numbers in the '4 times tables'. Label it F.

2. Within this set form a subset of numbers
in the '3 times tables'. Label it T.

3. Think of another subset that can be
formed. Remember to label your subset.

A. Decimals

What decimal part of each shape is coloured? The first one is done for you.

1.

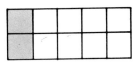

0·2 is coloured.

2.

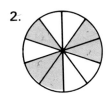

3.

4.

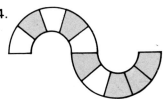

5. Write down the numbers which should be placed in the lettered boxes

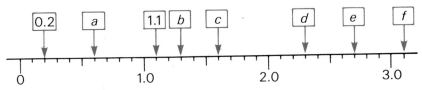

6. This sign > means 'greater than'. Use the > sign to say which number is the largest in each pair. The first one is done for you.

   a. 0·1, 1·5   **b.** 0·9, 0·6   **c.** 1·0, 0·5   **d.** 2·9, 3·1   **e.** 1·9, 2

   1·5 > 0·1

7. Re-arrange these numbers in order of the smallest to the biggest.

   a. 0·9, 0·7, 1·5   **b.** 1·7, 0·8, 2·5, 0·1   **c.** 0·2, 0, 2, 3·1, 1·7

B. Angles and shape

1. Copy the table on the left. Which angles are smaller than a right angle and which angles are larger than a right angle?

less than 90°	more than 90°

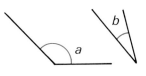
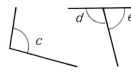
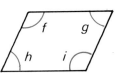

2. Copy and complete the table on the left. Find the order of rotation for each shape below.
   The order of rotation is the number of times a shape looks the same during one rotation.

Shape	Order of Rotation
a.	
b.	
c.	
d.	

a.

b.

c.

d.

C. Triangles

Find the size of the angles marked with letters.

1.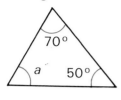

Angle a = ∗ °

2.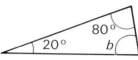

Angle b = ∗ °

3.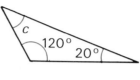

Angle c = ∗ °

4.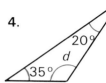

Angle d = ∗ °

Without measuring their angles, say whether these triangles are
a. Scalene   **b.** Isosceles   **c.** Equilateral

5.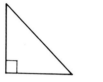

This is an____ triangle

6.

This is an____ triangle

7.

This is a ____ triangle

8.

This is an____ triangle

D. Fractions

Say what fraction of the whole has been coloured in each drawing.

1.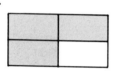

2.

3.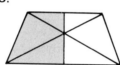

4.

5.

Complete the fractions under each drawing

6.

$\frac{3}{6} = \frac{1}{*}$

7.

$\frac{2}{8} = \frac{1}{*}$

8.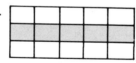

$\frac{*}{15} = \frac{1}{3}$

9.

$\frac{3}{9} = \frac{1}{*}$

10.

$\frac{9}{*} = \frac{*}{4}$

E. Coordinates

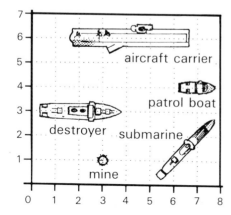

1. What would you hit at the coordinates (3, 1)?
2. What vessel would you hit at the coordinates (6, 1)?
3. Give one pair of coordinates to destroy the patrol boat.
4. Give the three pairs of coordinates to hit the destroyer.
5. Give the five pairs of coordinates to destroy the aircraft carrier.
6. Give the two pairs of coordinates to destroy the submarine.

F. Area

Find the area of the shapes below.

1.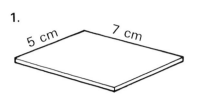
5 cm 7 cm

2. 4 cm
10 cm

3. 6 m
8 m

4.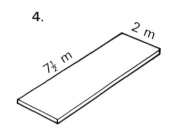
2 m
$7\frac{1}{2}$ m

Find the length of the sides marked with a question mark.

5. Area=50 cm² ?
10 cm

6. ?
Area =35 cm² 5 cm

7. Area = 100 m² ?
10 m

8. ?
Area = 9 cm²

9. ?
Area = 56 mm² 8 mm

Find the area of the triangles

10.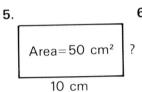
4 cm
12 cm

11.
12 cm 5 cm

12.
8 cm 10 cm

13. $4\frac{1}{2}$ cm 8 cm

14.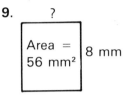
5 cm
9 cm

G. Distance

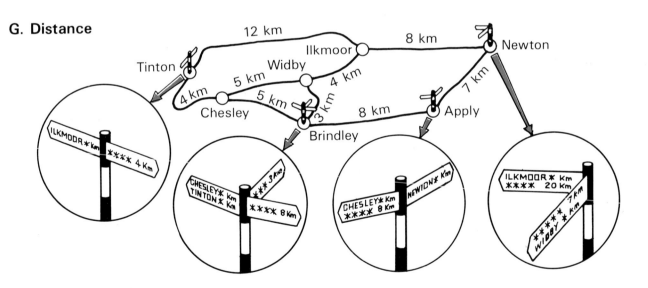

1. Copy and complete the village sign posts.
2. How far is it between Newton and Brindley, going by Apply?
3. How far is it from Chesley to Newton, going by the shortest route?
4. Janice walks at 5 km an hour. How many hours does it take her to walk from Tinton to Newton by the shortest route?

H. Number Patterns

Find the missing numbers on these factor trees.

1. 20 / 4, 5 / ?, ?

2. 36 / 9, 4 / ?, 3, 2, ?

3. 30 / 3, ? / 5, ?

4. 60 / 4, ? / ?, ?

5. Find all the factor pairs of the numbers below. The first one is done for you.

a. 12 = (12 x 1) = (3 x 4) = (6 x 2)

b. 30   **c.** 21   **d.** 45   **e.** 17

f. 40   **g.** 60   **h.** 19   **i.** 64

6. Which of the numbers above have only one factor pair? These numbers are called prime numbers.

7. Which numbers below are prime numbers?

a. 6, 4, 3, 9, 15   **b.** 1, 19, 8, 13, 21   **c.** 13, 9, 14, 23, 27, 31.

J. Time

1. Toby starts running at time A
He finishes running at time B

How long was he running for?

time A

time B

2. An airliner leaves Frankfurt at time X
It arrives at Heathrow at time Y

How long was the journey?

time X

time Y

3. From the time shown on this clock:—

a. add 20 minutes   **b.** add half an hour

c. subtract 3 hours   **d.** subtract 1 hour 5 minutes

4. 20.30   From the time shown on this clock:—

a. add 30 minutes   **b.** add 1 hour 25 minutes

c. subtract 2 hours   **d.** subtract 1 hour 15 minutes

5. Convert these 12-hour clock times to 24-hour clock times.

a. 11 a.m.   **b.** 8.30 p.m.   **c.** 11 p.m.   **d.** 1.25 p.m.   **e.** 1.15 a.m.

Section 5   **Area**

Exercise 1   Mr Green has been tiling walls in his house using square tiles. How many tiles can you see on each of these walls?

1.

2.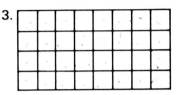

3.

4.

5.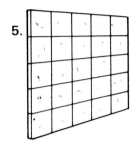

6.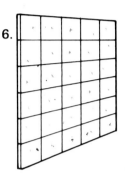

Exercise 2   For each drawing below copy and complete these two statements
a. The area of white tiles is ____ squares.
b. The area of coloured tiles is ____ squares.

1.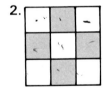

2.

3.

4.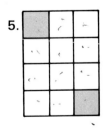

5.

6.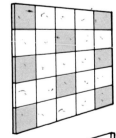

7.

8.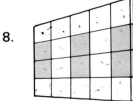

Exercise 3   None of these walls have been fully tiled. Calculate the total number of coloured tiles used when the jobs are completed.

1.    2. 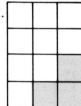   3.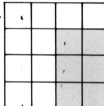

4.   5.    6.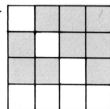

Exercise 4   If there are 24 tiles in each box, say which tile patterns below could be made from using exactly one full box of tiles.

1.    2. 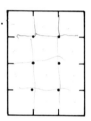   3.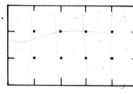

4.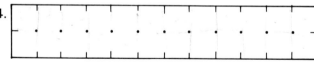

6. 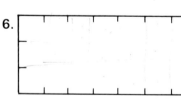   7.    5. (see right)

8.

9.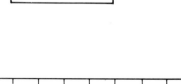

Using the centimetre-square

This is a centimetre-square. We write it as 1 cm². 1 cm We use this square to measure the area of small surfaces.
1 cm

Exercise 5

Copy and complete the two statements for each drawing below
a. The area of this shape is ____ cm x ____ cm
b. The area of this shape is ____ cm²
(The first one is done for you)

1. 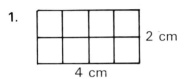 2 cm
4 cm

a. The area of this shape is 4 cm x 2 cm
b. The area of this shape is 8 cm²

2. 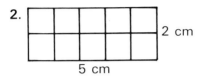 2 cm
5 cm

3. 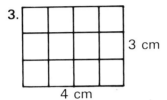 3 cm
4 cm

4. 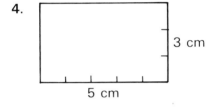 3 cm
5 cm

5. 6 cm
2 cm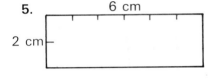

6. 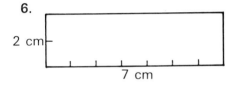 2 cm
7 cm

7.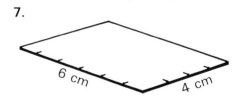
6 cm   4 cm

8.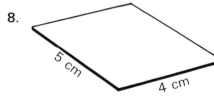
5 cm   4 cm

9.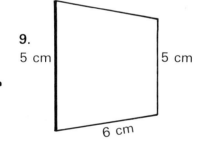
5 cm   5 cm
6 cm

Exercise 6

Copy and complete these three statements for each drawing below.
a. The area of A = ____ cm²
b. The area of B = ____ cm²
c. The total area of both A and B = ____ cm²

1.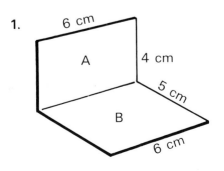
6 cm
A   4 cm
5 cm
B
6 cm

2.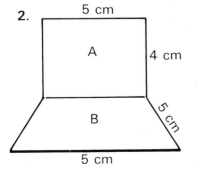
5 cm
A   4 cm
B   5 cm
5 cm

3.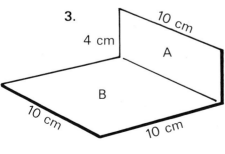
10 cm
4 cm
A
B
10 cm   10 cm

Surface area

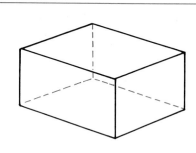

The surface area of this shape is the area of all the faces added together.

The dotted lines help us to imagine the three sides that we cannot see. This shape has 6 sides.

Exercise 7   Copy these drawings if you can. How many faces has each shape?

1.
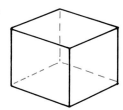

2.
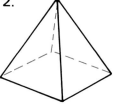

3.
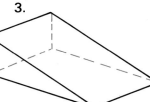

4.

5.
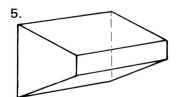

6.

Exercise 8   Find the surface area of each shape. The area of each face is printed on the sides of the shape.

1.

2.
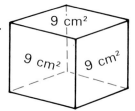

3.
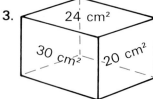

4.
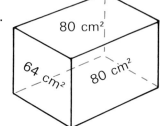

5.
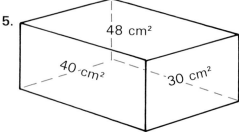

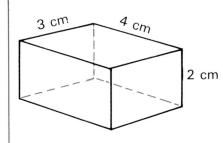

The surface area of this block is calculated by finding the area of all the six faces and adding the six areas together.

There are 3 pairs of sides **a.** top and bottom
                                      b. 2 long sides
                                        c. 2 short sides

Find the area of one side and you know the area of the opposite side.

Like this:

Area of top $= 3 \text{ cm} \times 4 \text{ cm}$
               $= 12 \text{ cm}^2$
Area of top and bottom $= 24 \text{ cm}^2$
Area of both short sides $= 6 \text{ cm}^2 + 6 \text{ cm}^2$
                         $= 12 \text{ cm}^2$

Area of both long sides $= 8 \text{ cm}^2 + 8 \text{ cm}^2$
                        $= 16 \text{ cm}^2$
Total surface area $= 24 \text{ cm}^2 + 16 \text{ cm}^2 + 12 \text{ cm}^2$
surface area $= 52 \text{ cm}^2$

Exercise 9

Find the surface area of each shape

1.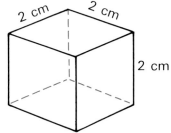

2.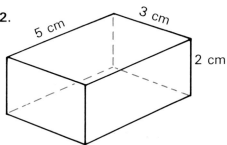

3.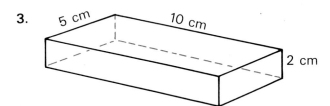

4.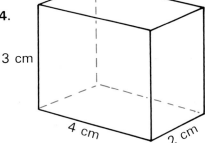

5.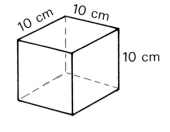

Section 6   The four rules

Addition and subtraction

Addition workcard 1

1. 206 +153	**2.** 317 +302
3. 660 +128	**4.** 235 +350
5. 129 +212	**6.** 228 +235
7. 547 +127	**8.** 416 +448

Addition workcard 2

1. 180 +453	**2.** 292 +161
3. 376 +482	**4.** 188 +350
5. 450 +196	**6.** 687 + 72
7. 66 +273	**8.** 658 + 80

Addition workcard 3

1. 288 +353	**2.** 167 +255
3. 296 +416	**4.** 426 +188
5. 678 +252	**6.** 199 +101
7. 1289 +3161	**8.** 2257 +5593

Addition workcard 4

1. 23 34 +22	**2.** 45 32 +12	**3.** 10 48 +21	**4.** 23 51 +21
5. 16 43 +22	**6.** 38 44 +14	**7.** 17 45 +23	**8.** 17 20 +39

Addition workcard 5

1. 48 19 +15	**2.** 27 17 +48	**3.** 18 37 +19	**4.** 49 17 +27
5. 59 5 +37	**6.** 28 27 + 8	**7.** 38 38 +14	**8.** 47 64 +29

Subtraction workcard 1

1. 80 −15	**2.** 70 −36
3. 60 −23	**4.** 72 −16
5. 51 −14	**6.** 34 −17
7. 76 −39	**8.** 85 −48

Subtraction workcard 2

1. 88 −15	**2.** 74 −53
3. 68 −25	**4.** 47 −23
5. 78 −38	**6.** 57 −55
7. 64 −63	**8.** 86 −56

Subtraction workcard 3

1. 272 −126	**2.** 580 −128
3. 853 −247	**4.** 681 −177
5. 634 −284	**6.** 513 −263
7. 705 −340	**8.** 636 −250

Subtraction workcard 4

1. 451 −246	**2.** 360 −255	**3.** 512 −305	**4.** 873 −267
5. 742 −156	**6.** 422 −236	**7.** 632 −274	**8.** 420 −156

Subtraction workcard 5

1. 613 −245	**2.** 440 −128	**3.** 710 −546	**4.** 550 −146
5. 702 −126	**6.** 504 −347	**7.** 601 −235	**8.** 502 −198

Multiplication

23 x 3

Multiply the units.

23 x
  3
___
  9

3 times 3 units is 9 units

Multiply the tens.

23 x
  3
___
69

3 times 2 tens is 6 tens

Exercise 1

Do these calculations

1. 21 x
       3

2. 24 x
       2

3. 32 x
       3

4. 13 x
       3

5. 43 x
       2

6. 12 x
       4

---

24 x 3

In some problems you will have to carry over numbers.

Multiply the units

24 x
  3
___
  2
___
  1

3 times 4 units is 12. 12 is 1 ten and 2 units. Write in the 2 units and carry the ten.

Multiply the tens

24 x
  3
___
72
___
  1

3 times 2 tens is 6 tens, plus the 1 ten carried = 7 tens.

Exercise 2

Do these calculations

1. 23 x
       4

2. 14 x
       4

3. 26 x
       3

4. 15 x
       5

5. 27 x
       2

6. 18 x
       3

7. 16 x
       5

8. 27 x
       3

9. 25 x
       3

10. 39 x
        2

11. 28 x
        3

12. 17x
        4

---

42 x 3

In some problems you will have to carry over from the tens column.

Multiply the units

42 x
  3
___
  6
___

3 times 2 units is 6 units.

Multiply the tens

42 x
  3
___
126
___

3 times 4 is 12 tens. 12 tens is 1 hundred and 2 tens. Write in the 2 tens and put the 1 hundred in the hundreds column.

Exercise 3

Do these calculations.

1. 32 x
       4

2. 43 x
       3

3. 52 x
       3

4. 61 x
       3

5. 41 x
       6

6. 52 x
       4

7. 41 x
       6

8. 51 x
       5

9. 43 x
       4

10. 54 x
        4

11. 46 x
        3

12. 36 x
        4

13. 124 x
        3

14. 215 x
        4

15. 106 x
        3

16. 217 x
        3

17. 162 x
        4

18. 256 x
        2

Long multiplication

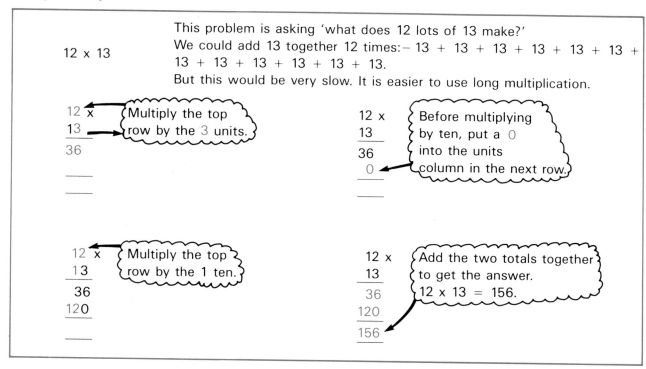

This problem is asking 'what does 12 lots of 13 make?'

We could add 13 together 12 times:– 13 + 13 + 13 + 13 + 13 + 13 + 13 + 13 + 13 + 13 + 13 + 13.

But this would be very slow. It is easier to use long multiplication.

12 × 13

Multiply the top row by the 3 units.

12 ×
13
36

Before multiplying by ten, put a 0 into the units column in the next row.

12 ×
13
36
0

Multiply the top row by the 1 ten.

12 ×
13
36
120

Add the two totals together to get the answer.
12 × 13 = 156.

12 ×
13
36
120
156

Exercise 4

Do these calculations. Look at the example above.

1. 21 ×
   12

2. 23 ×
   21

3. 12 ×
   23

4. 24 ×
   21

5. 13 ×
   23

6. 31 ×
   12

7. 122 ×
    32

8. 231 ×
    21

9. 121 ×
    42

10. 233 ×
    32

11. 314 ×
    22

12. 312 ×
    23

Copy and complete these workcards.
In these problems you will need to carry.

Long multiplication workcard 1

1. 214 ×
   23

2. 124 ×
   25

3. 231 ×
   14

3. 153 ×
   13

5. 263 ×
   31

6. 142 ×
   14

7. 126 ×
   23

8. 261 ×
   43

9. 143 ×
   34

Long multiplication workcard 2

1. 236 ×
   12

2. 126 ×
   31

3. 217 ×
   13

4. 261 ×
   41

5. 281 ×
   13

6. 173 ×
   31

7. 261 ×
   24

8. 247 ×
   22

9. 164 ×
   34

Division

Divide 2 into 4

$$2\overline{)482}$$

$$\frac{2}{2\overline{)482}}$$

2 goes into 4 twice. Put the answer 2 above the line.

Next divide 2 into 8

$$\frac{24}{2\overline{)4\,82}}$$

2 goes into 8 four times. Put the answer 4 above the line.

Then divide 2 into 2

$$\frac{241}{2\overline{)48\,2}}$$

2 goes into 2 once. Put the answer 1 above the line.

482 divided by 2 is 241

Exercise 5

1. $2\overline{)264}$ 2. $3\overline{)639}$ 3. $2\overline{)428}$ 4. $3\overline{)366}$ 5. $2\overline{)862}$ 6. $3\overline{)363}$

7. $4\overline{)488}$ 8. $2\overline{)848}$ 9. $3\overline{)969}$ 10. $2\overline{)886}$ 11. $4\overline{)888}$ 12. $3\overline{)699}$

Divide 2 into 6

$$2\overline{)612}$$

$$\frac{3}{2\overline{)612}}$$

2 goes into 6 three times. Put the answer 3 above the line.

Divide 2 into 1

$$\frac{30}{2\overline{)6\,12}}$$

2 does not divide into 1, so as to give a whole number. Put 0 above the line.

Divide 2 into 12

$$\frac{30\,6}{2\overline{)6\,12}}$$

2 did not divide into 1. So divide 2 into 12. It divides 6 times. Put the answer above the line.

Division workcard 1

1. $2\overline{)412}$ 2. $2\overline{)214}$

3. $3\overline{)612}$ 4. $3\overline{)315}$

5. $4\overline{)816}$ 6. $5\overline{)515}$

7. $2\overline{)128}$ 8. $3\overline{)129}$

Division workcard 2

1. $2\overline{)2184}$ 2. $2\overline{)4612}$

3. $3\overline{)6129}$ 4. $3\overline{)3186}$

5. $3\overline{)3915}$ 6. $5\overline{)5105}$

7. $3\overline{)6216}$ 8. $4\overline{)4248}$

Division workcard 3

1. $4\overline{)424}$ 2. $3\overline{)921}$

3. $5\overline{)525}$ 4. $5\overline{)530}$

5. $4\overline{)820}$ 6. $3\overline{)624}$

7. $6\overline{)624}$ 8. $4\overline{)432}$

Exercise 6

These problems will leave you with a remainder.

$$\frac{12}{2\overline{)24\,5}}$$

When you divide 2 into 5 it will divide twice and you will have 1 left over.

$$\frac{12\,2\text{ r }1}{2\overline{)245}}$$

1. $2\overline{)487}$ 2. $2\overline{)625}$ 3. $3\overline{)397}$ 4. $3\overline{)665}$ 5. $4\overline{)847}$ 6. $3\overline{)638}$

7. $2\overline{)325}$ 8. $3\overline{)757}$ 9. $4\overline{)853}$ 10. $3\overline{)646}$ 11. $2\overline{)713}$ 12. $3\overline{)791}$

Exercise 7

1. If 214 nails are shared between 2 boxes, how many nails are there in each box?

2. Four people share a prize of £824. How much do they each get?

3. If 225 soldiers are divided into 5 groups, how many are in each group?

4. In 4 months Mrs Hart saved £1216. How much did she save each month?

5. A rope 245 m long is to be cut up into 5 m lengths. How many 5 m lengths will there be?

$12\overline{)156}$

$\begin{array}{r} 1 \\ 12\overline{)156} \end{array}$

Does 12 go into 1? No, it does not. Does 12 go into 15? It goes once, with 3 left over.

$\begin{array}{r} 1 \\ 12\overline{)15^36} \end{array}$

Carry the 3 over to the next figure. This makes 36.

$\begin{array}{r} 13 \\ 12\overline{)15^36} \end{array}$

Does 12 go into 36? It goes exactly 3 times. 156 divided by 12 is 13.

Exercise 8

1. $13\overline{)286}$ 2. $12\overline{)252}$ 3. $10\overline{)230}$ 4. $13\overline{)273}$ 5. $11\overline{)132}$

6. $12\overline{)156}$ 7. $14\overline{)168}$ 8. $10\overline{)140}$ 9. $12\overline{)264}$ 10. $14\overline{)322}$

11. $15\overline{)315}$ 12. $15\overline{)195}$ 13. $14\overline{)294}$ 14. $12\overline{)168}$ 15. $16\overline{)368}$

Division workcard 4	
1. $12\overline{)290}$	2. $15\overline{)157}$
3. $13\overline{)306}$	4. $14\overline{)202}$
5. $11\overline{)117}$	6. $14\overline{)329}$
7. $17\overline{)376}$	8. $16\overline{)213}$

In each of these problems there is a remainder.

Division workcard 5	
1. $14\overline{)1442}$	2. $11\overline{)1331}$
3. $16\overline{)1808}$	4. $15\overline{)1980}$
5. $12\overline{)2424}$	6. $15\overline{)3165}$
7. $13\overline{)3003}$	8. $17\overline{)2074}$

Division workcard 6	
1. $13\overline{)1735}$	2. $15\overline{)1570}$
3. $21\overline{)2778}$	4. $22\overline{)4716}$
5. $18\overline{)4166}$	6. $23\overline{)3108}$
7. $32\overline{)4230}$	8. $25\overline{)5288}$

In each of these problems there is a remainder.

Brackets

In sums like 2 + 4 x 3 where there is more than one calculation to do, the order in which you add and multiply is important.

The problem 2 + 4 x 3 can give you two correct answers

2 add 4 is 6; 6 times 3 is 18
or
4 times 3 is 12; 12 add 2 is 14

We put in brackets to show which calculation to do first.
Do the calculation you see in the brackets first.

(2 + 4) x 3 = 6 x 3   (4 x 3) + 2 = 12 + 2
           = 18   = 14

Exercise 9

Find the value of each bracket then add 5 to each value

1. (10 x 2) + 5   **2.** (15 ÷ 5) + 5   **3.** (7 x 5) + 5
4. (56 ÷ 8) + 5   **5.** (16 x 2) + 5   **6.** (17 x 3) + 5

Find the value of each bracket then multiply each value by 2.

7. (10 + 5) x 2   **8.** (7 + 13) x 2   **9.** (29 − 15) x 2
10. (15 − 3) x 2   **11.** (16 + 5) x 2   **12.** (52 − 22) x 2

Exercise 10

Work out the value of these calculations. Remember that you must do the calculation in the brackets first.

1. (3 + 7) ÷ 2   **2.** (5 x 3) + 12   **3.** (15 x 2) − 9
4. (20 x 5) + 14   **5.** (16 ÷ 2) + 13   **6.** (28 ÷ 4) − 2
7. 6 + (5 x 7)   **8.** 14 − (30 ÷ 6)   **9.** 25 x (17 − 15)

Exercise 11

Copy out the calculations and answers below. The brackets have been left out. Put the brackets onto the calculations to produce the answers shown.

1. 6 + 3 x 2 = 18   **2.** 5 x 3 − 12 = 3   **3.** 7 x 2 + 1 = 15
4. 5 + 7 x 2 = 19   **5.** 15 ÷ 10 − 7 = 5   **6.** 9 x 3 + 2 = 29
7. 14 + 1 x 3 = 17   **8.** 16 ÷ 4 + 4 = 2   **9.** 20 ÷ 27 − 23 = 5

Rectangular block

Sphere

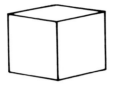
Cube

Pyramid

Cylinder

Cone

Exercise 1

1. 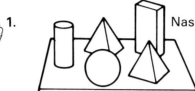 Nasty Eric knocked all these solids off the table.

Which drawing below shows the mess Eric made?

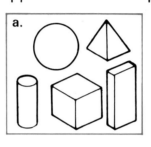

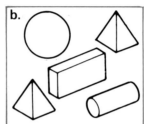

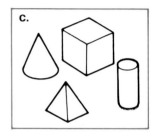

2. Eric now knocks over these solids. Which drawing below shows the mess Eric made?

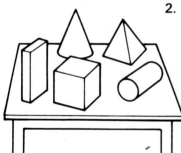

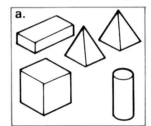

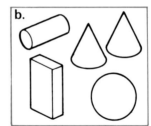

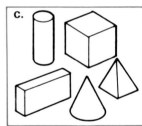

Exercise 2

Copy and complete these sentences.

1. There are ＿＿ spheres in this drawing.

2. There are ＿＿ cubes in this drawing.

3. There are ＿＿ cones in this drawing.

4. There are ＿＿ cylinders in this drawing.

5. There are ＿＿ pyramids in this drawing.

6. There are ＿＿ rectangular blocks in this drawing.

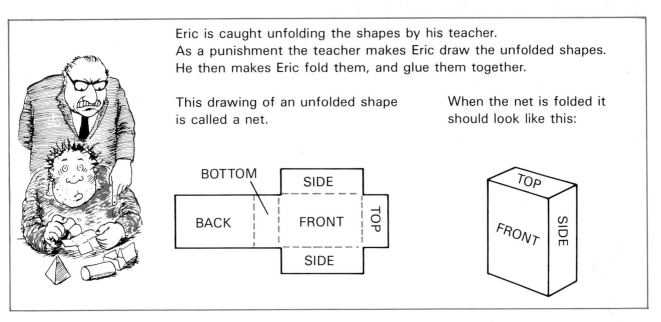

Eric is caught unfolding the shapes by his teacher.
As a punishment the teacher makes Eric draw the unfolded shapes.
He then makes Eric fold them, and glue them together.

This drawing of an unfolded shape is called a net.

When the net is folded it should look like this:

Exercise 3

Here are the nets for a cube and a pyramid. Carefully draw or trace them onto card. Fold the nets and glue them to make solid shapes. Notice that you must leave tabs so that you can glue the edges together.

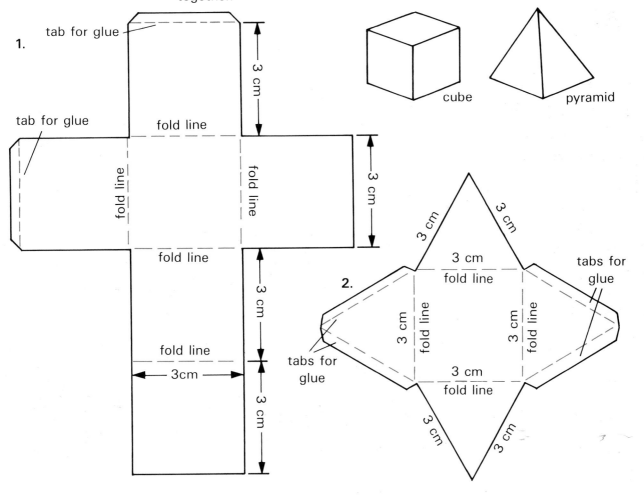

Eric's class have made some shapes out of cardboard.

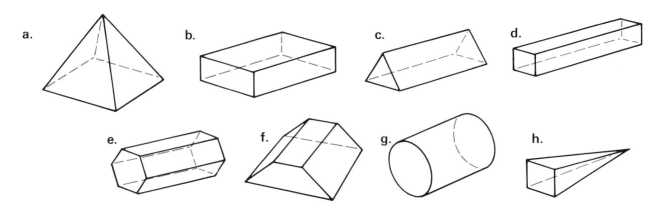

a. b. c. d.

e. f. g. h.

Exercise 4

Eric is in a bad mood. He pulls apart all the shapes.
When he has unfolded the shapes they look like these below.

Which shape above do you think matches up with the unfolded shape below?

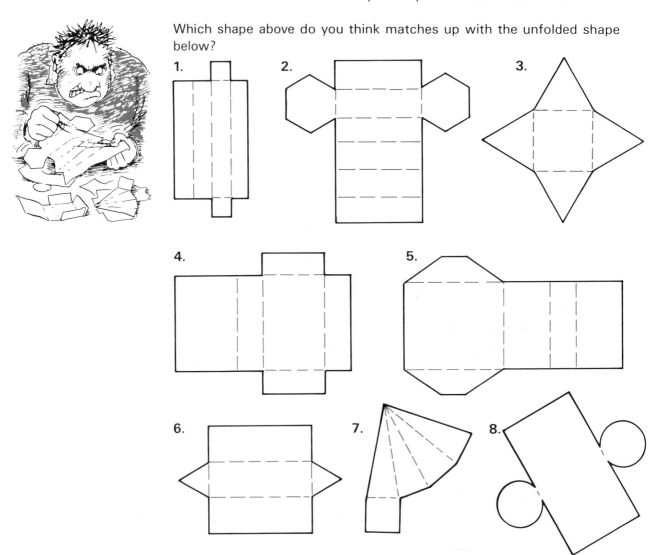

1.   2.   3.

4.   5.

6.   7.   8.

For his homework Eric is told to draw the net for this shape.

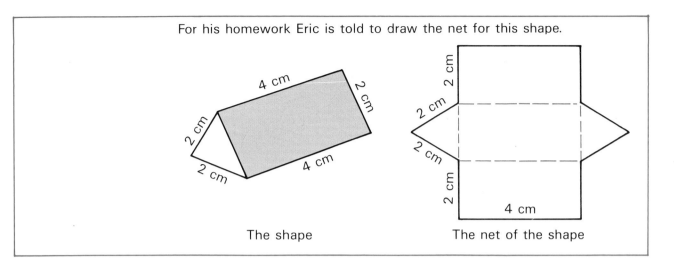

The shape

The net of the shape

Exercise 5

Draw the nets for these shapes.
Draw the nets full size.

1.

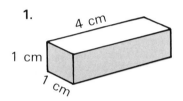

2.

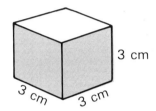

Exercise 6

1. How many spheres are there in this puppet?
2. How many cones are there in this puppet?
3. How many cylinders are there in this puppet?
4. How many rectangular blocks are there in this puppet?
5. Which one of these drawings shows all the pieces of the puppet when it is taken apart?

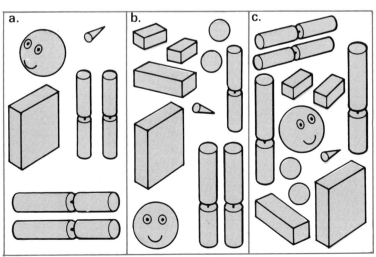

a.

b.

c.

Volume

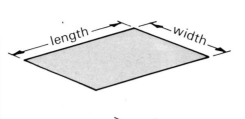

This shape has two measurements: length and width.

Shapes or surfaces with only two measurements only have area.

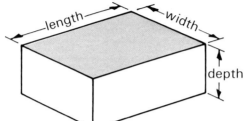

This shape has three measurements length, width and depth.

Shapes with three measurements have volume.

Area measures how much space is taken up by a flat surface.
Volume measures how much space is taken up by a solid shape.
Solid shapes have three measurements.

Exercise 7

Look at the drawings below. Copy and complete the table below. Make a list of the solids and make a list of the flat surfaces.

surfaces	solids
a	b

a.

b.

c.

d.

e.

f.

g.

h.

i.

j.

k.

l.

m.

n.

This is a cube.
Its sides are the same length.

Cubes can be used to measurement volume.

The volume of this block is 6 cubes.

Exercise 8

Find the volume of each block or shape.

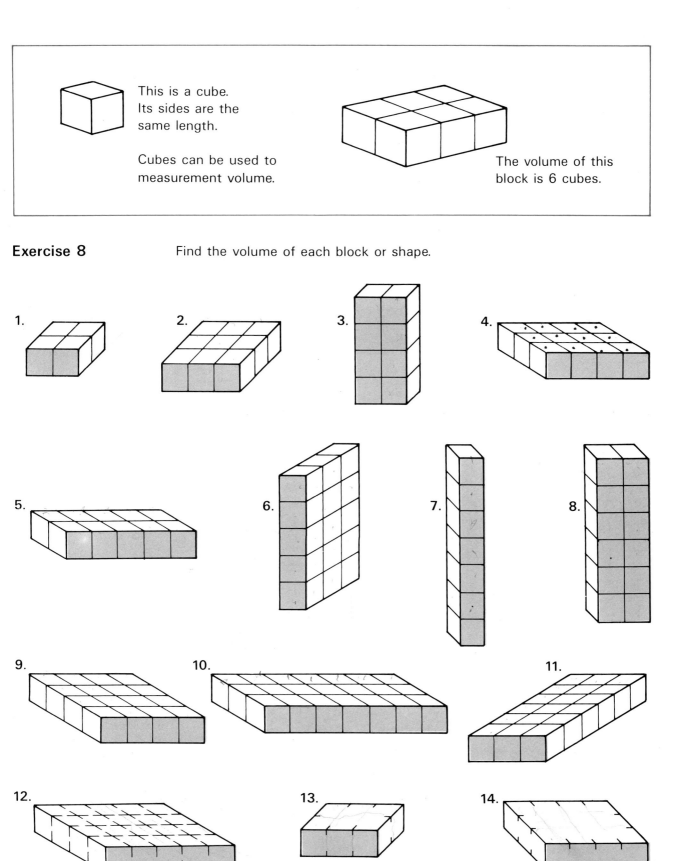

1.

2.

3.

4.

5.

6.

7.

8.

9.

10.

11.

12.

13.

14.

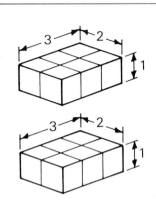

Each of these layers has
a volume of 6 cubes.

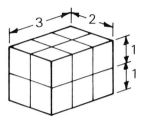

When the two blocks are put together
they have a total volume of 12 cubes.

We can find volume without counting cubes.
Find the number of cubes in the top layer then multiply by the number
of layers in the block or multiply length x width x height.

Exercise 9   Copy and complete the sentences below.

1.

The volume of the block
is ____ cubes

2.

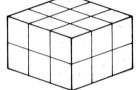

The volume of the block
is ____ cubes

3.

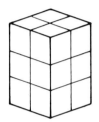

The volume of the block
is ____ cubes

4.

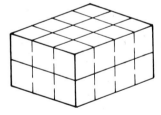

The volume of the block
is ____ cubes.

5.

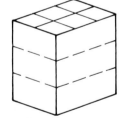

The volume of the block
is ____ cubes.

6.

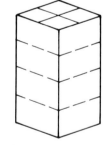

The volume of the block
is ____ cubes.

7.

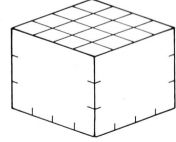

The volume of the block
is ____ cubes.

8.

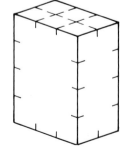

The volume of the block
is ____ cubes.

9.

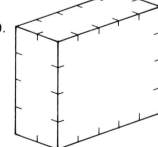

The volume of the block
is ____ cubes.

Standard cubes

If we use cubes as a way of measuring volume, we need to make sure that we are all using the same size of cube to find volume.

We use standard size cubes. This drawing shows a cube. All of its sides are 1 cm long and it is called a centimetre-cube (written 1 cm³)

The centimetre-cube is used for measuring small volumes.

Exercise 10 Find the volume of these objects and complete the sentences below. The drawings are not full size.

1.

The volume of the box is 4 cm x 3 cm x 1 cm The volume is * cm³

2.

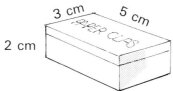

The volume of the box is 3 cm x 5 cm x 2 cm. The volume is * cm³

3.

The volume of this box is 2 cm x 4 cm x 5 cm. The volume is * cm³

4.

The volume is * cm³

5.

The volume is * cm³

6.

The volume is * cm³

7.

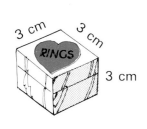

The volume is * cm³

8.

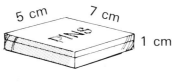

The volume is * cm³

9.

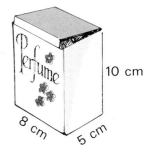

The volume is * cm³

Exercise 11   Find the volume of each block. The shapes are not drawn to scale.

1.

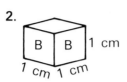

Block A has a volume
of * cm³

2.

Block B has a volume
of * cm³

3.

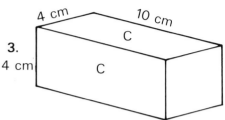

Block C has a volume
of * cm³

4.

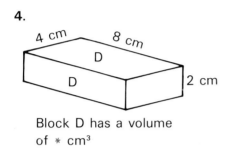

Block D has a volume
of * cm³

5.

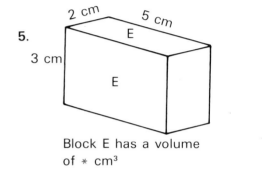

Block E has a volume
of * cm³

Exercise 12   Find the volume of the shapes below. Each shape is made up from the blocks above.

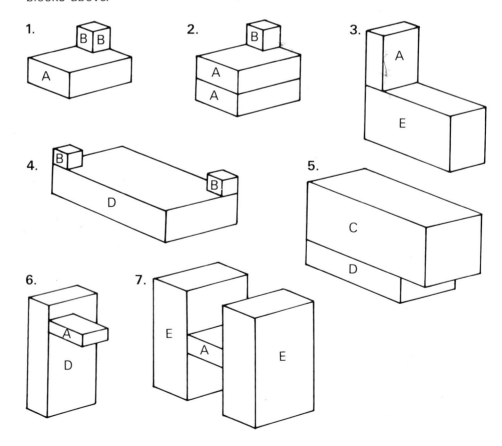

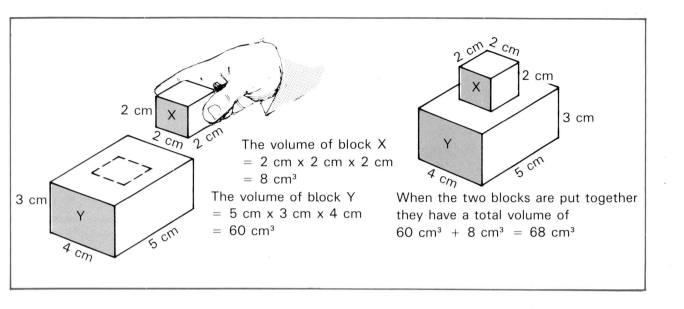

The volume of block X
= 2 cm x 2 cm x 2 cm
= 8 cm³

The volume of block Y
= 5 cm x 3 cm x 4 cm
= 60 cm³

When the two blocks are put together
they have a total volume of
60 cm³ + 8 cm³ = 68 cm³

Exercise 13   Each shape below is made from more than one block. The dotted lines show where the blocks are joined. Find the total volume of each shape.

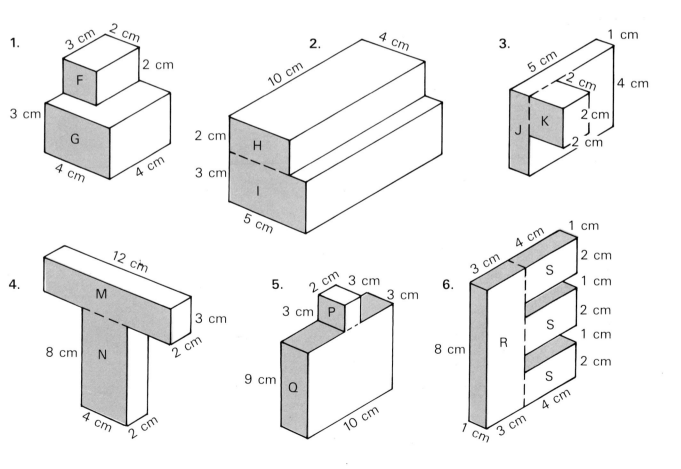

Section 8   Minus numbers

This is a thermometer. It is used to measure temperature.
0° is called freezing point. At this temperature water freezes and becomes ice.
Temperature is measured in degrees Celsius or °C.

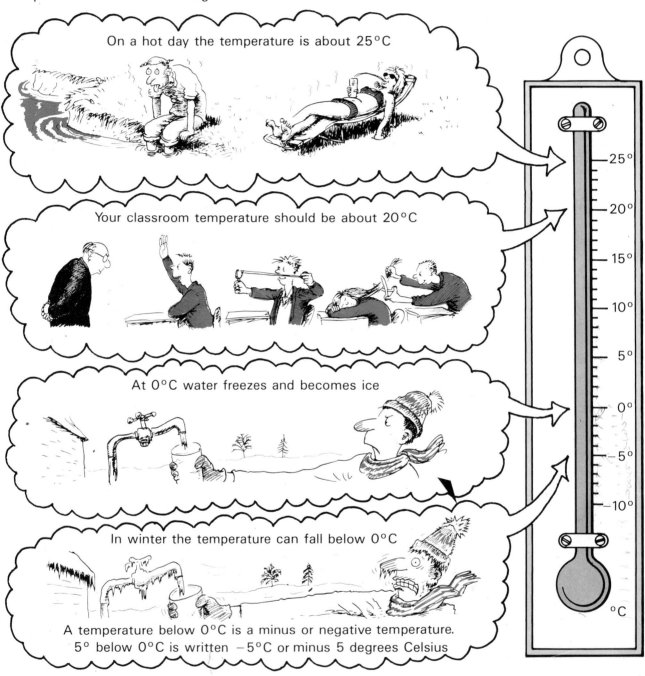

On a hot day the temperature is about 25°C

Your classroom temperature should be about 20°C

At 0°C water freezes and becomes ice

In winter the temperature can fall below 0°C

A temperature below 0°C is a minus or negative temperature.
5° below 0°C is written −5°C or minus 5 degrees Celsius

Exercise 1   Which of the temperatures are below 0°C

1. −2°C   **2.** −4°C   **3.** 9°C   **4.** 11°C   **5.** −1°C

6. 29°C   **7.** −30°C   **8.** −22°C   **9.** 121°C   **10.** −130°C

Exercise 2

Rearrange the temperatures in each question from the coldest to the hottest.
Use the thermometer on the previous page to help you.

1. 1°C, 5°C, 0°C

2. 1°C, 4°C, 6°C

3. 0°C, −1°C, 3°C

4. 1°C, 0°C, −2°C

5. −1°C, −5°C, −2°C

6. −5°C, −10°C, 5°C

7. −10°C, −8°C, −20°C

8. −18°C, 9°C, 12°C, −15°C

9. −7°C, 0°C, −6°C, −11°C

10. 9°C, −1°C, −8°C, 3°C

11. −1°C, −14°C, −3°C, −12°C

12. −5°C, 13°C, −21°C, −1°C

Rising temperature
The temperature starts at −1°C.

It rises by 5° and
finishes at 4°C.

When the temperature rises we add,

so we can write $(-1) + 5 = 4$

$+5$

```
5°
4°
3°
2°
1°
0°
-1°
-2°
-3°
°C
```

The temperature starts at −6°C
it rises by 4° and
finishes at −2°C.

So we write

$(-6) + 4 = (-2)$

$+4$

```
2°
1°
0°
-1°
-2°
-3°
-4°
-5°
-6°
°C
```

Exercise 3

Copy and complete the table.

Starting temp	Rise in temp	Finishing temp
0°C	8°	*°C
−2°C	2°	*°C
−9°C	7°	*°C
−7°C	12°	*°C
−11°C	16°	*°C
−10°C	7°	*°C
−12°C	20°	*°C
−18°C	9°	*°C
−13°C	5°	*°C
−10°C	15°	*°C

Exercise 4

Copy and complete the calculations. The thermometer will help you.

1. $0 + 7 = *$

2. $(-1) + 5 = *$

3. $(-2) + 6 = *$

4. $(-5) + 5 = *$

5. $(-1) + 1 = *$

6. $(-8) + 10 = *$

7. $(-4) + 12 = *$

8. $(-6) + 15 = *$

9. $(-10) + 7 = *$

10. $(-13) + 8 = *$

11. $(-15) + 9 = *$

12. $(-14) + 7 = *$

```
15°

10°

5°

0°

-5°

-10°
°C.
```

Falling temperature

The temperature starts at 3°C. It falls by 6° and finishes at −3°C.

When the temperature falls we subtract.

So we write 3 − 6 = (−3)

°C
┌ 4°
├ 3°
├ 2°
├ 1°
├ 0°
├ −1°
├ −2°
├ −3°
├ −4°

−6

The temperature on this thermometer starts at −1°C and falls by 5°. It finishes at −6°C.

So we write (−1) − 5 = (−6)

°C
┌ 1°
├ 0°
├ −1°
├ −2°
├ −3°
├ −4°
├ −5°
├ −6°
├ −7°

−5

Exercise 5   Copy and complete the table.

Starting temp	Fall in temp	Finishing temp
10°C	5°	* °C
5°C	6°	* °C
7°C	10°	* °C
0°C	8°	* °C
−2°C	8°	* °C
−4°C	6°	* °C
−6°C	5°	* °C
−9°C	2°	* °C
15°C	25°	* °C
−9°C	11°	* °C

Exercise 6   Copy and complete the calculations below.

1. 10 − 7 = *   2. 7 − 0 = *   3. 0 − 5 = *

4. 0 − 10 = *   5. 1 − 3 = *   6. 2 − 6 = *

7. (−1) − 4 = *   8. (−5) − 5 = *   9. (−2) − 9 = *

10. (−8) − 10 = *   11. 3 − 12 = *   12. 7 − 13 = *

13. (−6) − 14 = *   14. (−5) − 11 = *   15. (−20) − 5 = *

Exercise 7   Rewrite these sentences as simple calculations. The first one is done for you.

1. The temperature started at −2°C and rose by 5° to 3°C (−2) + 5 = 3

2. The temperature rose by 7° from −3°C to 4°C (start (−3) + . . .)

3. Overnight the temperature fell by 10° from 8°C to −2°C (start 8 − . . .)

4. This morning the temperature started at −7°C. It rose by 9° and finished at 2°C (start (−7) . . .)

5. Today's temperature is −5°C. Tomorrow the temperature will rise by 11° to 6°C (start (−5) . . .)

Blunderman

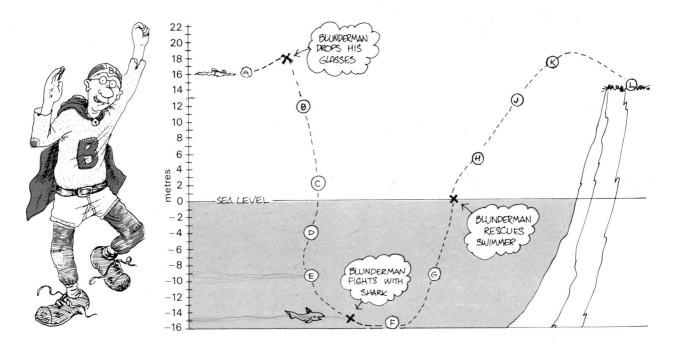

Exercise 8

Blunderman can fly through the air and under water. He starts his journey at point A.

1. How high above sea-level is he at point A?

2. How high above sea-level is he when he drops his glasses?

3. How high above sea-level is he at point B?

4. How many metres does he descend between points B and C?

5. How far below sea-level is he at point D?

6. How far below sea-level is he at point E?

7. How far below sea-level did Blunderman fight the shark?

8. How far below sea-level is he at point F?

9. What does Blunderman do at a point zero metres above sea-level?

10. What letters will you find on the (−9) metre level?

11. How high above sea-level is point L?

12. How many metres does Blunderman ascend between points J and K?

13. How many metres does he ascend between points H and J?

14. How many metres does he ascend between points F and G?

15. How many metres does he descend between points B and F?

Section 9 Fractions

The pupils at Dimwig Hall School are in for some bad news.
Due to a shortage of money, Dr Dimwig, the head, has to
announce cut backs.

*AS FROM TOMORROW, ALL FOOD RATIONS WILL BE CUT BY **HALF**!*

Before the cut backs the pupils' plates looked like this .

BREAKFAST

LUNCH

DINNER

Exercise 1

Re-draw these plates showing the food rations cut by half.

> To find half of a number divide by 2.
> To find half of 8 we divide 8 by 2, 8 ÷ 2 = 4

Exercise 2

In their maths lessons the pupils have to practise finding halves of
amounts.
Find a half of each of these amounts.

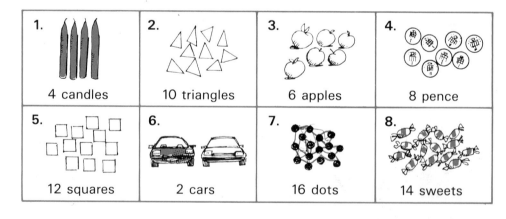

| 1. 4 candles | 2. 10 triangles | 3. 6 apples | 4. 8 pence |
| 5. 12 squares | 6. 2 cars | 7. 16 dots | 8. 14 sweets |

Exercise 3

1. There are 20 people on a bus. Half of them get off. How many are left?
2. Ali has 24 comics. He gives half of them away. How many has he left?
3. Mandy has 30p. She spends half of it. How much has she left?
4. Roger ate half of his sweets. He has 8 sweets left. How many did he start with?

Classes are divided into four house groups. There are 12 pupils in class 3Y. So there are 3 pupils in each house. $\frac{1}{4}$ of 12 = 3. To find $\frac{1}{4}$ of a number we divide by 4.

Exercise 4

Look at the pictures below and say how many pupils will belong in each house. Remember there are four houses.

1.

Class 1M

2.

Class 2R

3.

Class 3H

4.

Class 4P

Exercise 5

Calculate $\frac{1}{4}$ of the number of crosses in each box below.

1.
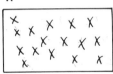

2.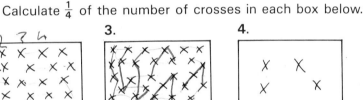

3.

4.
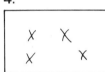

5.

Answer the questions below. The first one is done for you.

6. $\frac{1}{4}$ of 12 = 3   7. $\frac{1}{4}$ of 8 = *   8. $\frac{1}{4}$ of 20 = *   9. $\frac{1}{4}$ of 16 = *

10. $\frac{1}{4}$ of 40 = *   11. $\frac{1}{4}$ of 32 = *   12. $\frac{1}{4}$ of 80 = *   13. $\frac{1}{4}$ of 100 = *

To find $\frac{3}{4}$ of 8:

Find, $\frac{1}{4}$ of 8
$\frac{1}{4}$ of 8 = 2

If $\frac{1}{4}$ of 8 = 2
$\frac{3}{4}$ of 8 = 3 x 2
$\frac{3}{4}$ of 8 = 6

Exercise 6

Answer the questions below. The first one has been done for you.

1. $\frac{3}{4}$ of 8 = 6   2. $\frac{3}{4}$ of 4 = *   3. $\frac{3}{4}$ of 20 = *   4. $\frac{3}{4}$ of 12 = *

5. $\frac{3}{4}$ of 16 = *   6. $\frac{3}{4}$ of 24 = *   7. $\frac{3}{4}$ of 32 = *   8. $\frac{3}{4}$ of 40 = *

9. $\frac{3}{4}$ of 80 = *   10. $\frac{3}{4}$ of 60 = *   11. $\frac{3}{4}$ of 44 = *   12. $\frac{3}{4}$ of 100 = *

The pupils at Dimwig Hall have been told they must give $\frac{1}{3}$ of their pocket money to the school fund. To find $\frac{1}{3}$ of their pocket money the pupils divide their money by 3.

Exercise 7

Each pupil in the table below gave $\frac{1}{3}$ of their money to the school fund. Copy and complete the table showing how much each pupil gave and how much each kept.

Name	Pocket Money	$\frac{1}{3}$ of pocket money	Money kept
Martin	£15	£5	£10
Ali	£9		
Betty	£18		
Rachid	£21		
Ray	£3		
Derek	£30		
Arun	£6		
Keran	£12		
Dinish	£33		
Rosie	£45		
Don	£60		

Exercise 8

Answer the questions below. The first one is done for you.

1. $\frac{1}{3}$ of 9 = 3   **2.** $\frac{1}{3}$ of 3 = *   **3.** $\frac{1}{3}$ of 30 = *   **4.** $\frac{1}{3}$ of 15 = *

5. $\frac{1}{3}$ of 12 = *   **6.** $\frac{1}{3}$ of 18 = *   **7.** $\frac{1}{3}$ of 60 = *   **8.** $\frac{1}{3}$ of 90 = *

£9

To find $\frac{2}{3}$ we divide by 3 and multiply by 2

$\frac{1}{3}$ of £9 = £3

$\frac{2}{3}$ of £9 = £6

Exercise 9

Answer the questions below. The first one is done for you.

1. $\frac{2}{3}$ of 9 = 6   **2.** $\frac{2}{3}$ of 12 = *   **3.** $\frac{2}{3}$ of 15 = *   **4.** $\frac{2}{3}$ of 30 = *

5. $\frac{2}{3}$ of 18 = *   **6.** $\frac{2}{3}$ of 24 = *   **7.** $\frac{2}{3}$ of 36 = *   **8.** $\frac{2}{3}$ of 90 = *

Good news at Dimwig Hall! Dr Dimwig has decided that holidays will be increased by $\frac{1}{5}$.

To find $\frac{1}{5}$ of an amount we divide it by 5.

$\frac{1}{5}$ of 10 is $10 \div 5 = 2$

$\frac{1}{5}$ of 10 = 2

Exercise 10

Answer the questions below. The first one is done for you.

1. $\frac{1}{5}$ of 10 = 2

2. $\frac{1}{5}$ of 20 = *

3. $\frac{1}{5}$ of 5 = *

4. $\frac{1}{5}$ of 15 = *

5. $\frac{1}{5}$ of 25 = *

6. $\frac{1}{5}$ of 35 = *

7. $\frac{1}{5}$ of 50 = *

8. $\frac{1}{5}$ of 60 = *

9. $\frac{1}{5}$ of 40 = *

10. $\frac{1}{5}$ of 55 = *

11. $\frac{1}{5}$ of 75 = *

12. $\frac{1}{5}$ of 100 = *

Exercise 11

This table shows the length of school holidays at Dimwig Hall.

a. Find $\frac{1}{5}$ of each holiday in days

b. Increase each holiday by $\frac{1}{5}$

c. Copy and complete the table.

Holidays	Summer	Autumn	Christmas	Spring	Easter	Whitsun
Number of days	35	5	20	10	15	5
$\frac{1}{5}$ increase in days	*	*	*	2	*	1
New number of days in each holiday	*	*	*	12	*	6

To find $\frac{3}{5}$ of 10 we start by finding $\frac{1}{5}$

$\frac{1}{5}$ of 10 = $10 \div 5 = 2$

$\frac{3}{5}$ of 10 = $3 \times 2 = 6$

Exercise 12

1. Calculate $\frac{3}{5}$ of these amounts

    a. 10   b. 15   c. 5   d. 20   e. 30   f. 25   g. 40

2. Calculate $\frac{2}{5}$ of the amounts in question 1.

3. Calculate $\frac{4}{5}$ of the amounts in question 1.

Exercise 13   Complete the sentences below.

1.

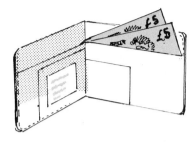

$\frac{1}{5}$ of the money in the wallet can be seen. The wallet has £ * in it.

2.

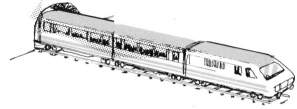

$\frac{1}{4}$ of the train can be seen. Altogether the train has * coaches

3.

You can see $\frac{2}{5}$ of the biscuits. Altogether the box holds * biscuits

4.

The box holds 40 matches. The fraction taken from the box is *

5.

The fraction of the pile that is missing is *

6.

You can see $\frac{2}{5}$ of the windows of the tower. The tower has * windows altogether.

7.

$\frac{3}{4}$ of the books on the shelf are shown. There should be * books altogether.

8.

$\frac{3}{4}$ of the cakes have been eaten. There were * cakes before any were eaten.

Exercise 14   Answer the questions below.

1. Pat has spent $\frac{2}{3}$ of her money. She started with £18. How much is left?

2. How much is $\frac{2}{5}$ of £1·50?

3. How much is $\frac{2}{3}$ of 300 kg?

4. How many years are there in $\frac{4}{5}$ of a century?

5. How many minutes are there in $\frac{2}{3}$ of 2 hours?

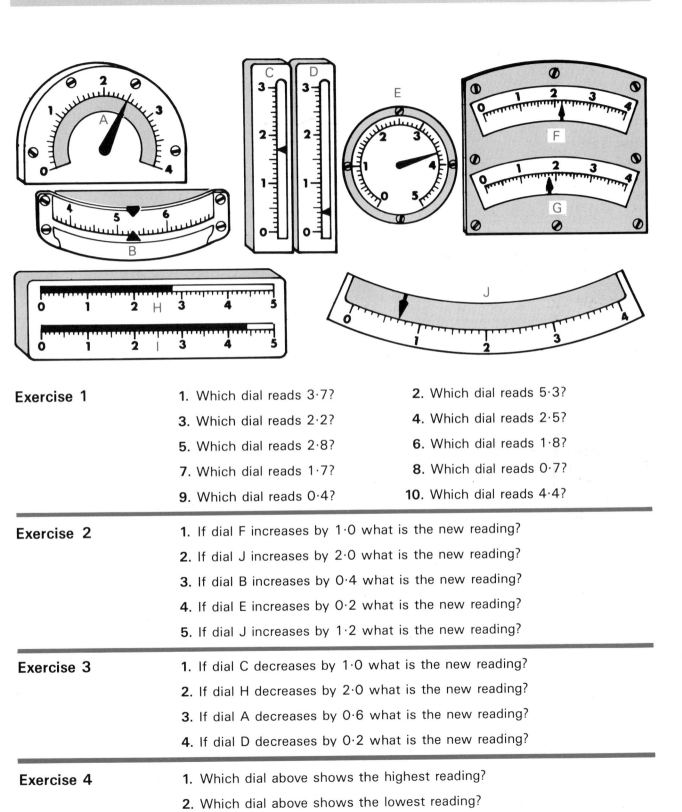

Exercise 1

1. Which dial reads 3·7?
2. Which dial reads 5·3?
3. Which dial reads 2·2?
4. Which dial reads 2·5?
5. Which dial reads 2·8?
6. Which dial reads 1·8?
7. Which dial reads 1·7?
8. Which dial reads 0·7?
9. Which dial reads 0·4?
10. Which dial reads 4·4?

Exercise 2

1. If dial F increases by 1·0 what is the new reading?
2. If dial J increases by 2·0 what is the new reading?
3. If dial B increases by 0·4 what is the new reading?
4. If dial E increases by 0·2 what is the new reading?
5. If dial J increases by 1·2 what is the new reading?

Exercise 3

1. If dial C decreases by 1·0 what is the new reading?
2. If dial H decreases by 2·0 what is the new reading?
3. If dial A decreases by 0·6 what is the new reading?
4. If dial D decreases by 0·2 what is the new reading?

Exercise 4

1. Which dial above shows the highest reading?
2. Which dial above shows the lowest reading?
3. Which dials show readings between: **a.** 2·0 and 3·0? **b.** 3·0 and 5·0?

Addition

Remember to keep the numbers in the correct column when adding.

3·3	3·3
13·4	13·4
+ 5·2	+ 5·2
CORRECT	WRONG

Make sure you keep the decimal points in line.

Exercise 5

Workcard 1

1. 2·4
   + 1·3

2. 4·5
   + 3·0

3. 5·5
   + 2·3

4. 6·0
   + 3·8

5. 3·4
   +12·3

6. 2·1
   +14·6

7. 16·7
   +12·2

8. 21·5
   +14·4

Workcard 2

1. 4·6
   + 2·5

2. 2·8
   + 5·5

3. 4·9
   + 3·2

4. 3·7
   + 3·7

5. 10·7
   + 3·4

6. 0·9
   +15·5

7. 16·7
   + 0·8

8. 22·9
   + 7·6

Workcard 3

1. 26·8
   +30·9

2. 20·7
   + 19·7

3. 19·7
   + 3·3

4. 0·9
   + 19·3

5. 26·9
   + 17·7

6. 35·5
   + 6·6

7. 10·7
   +39·3

8. 29·6
   + 10·4

Exercise 6

1.
   2.4
   3.1
   + 14.3
   ―――
   19.8

Add up these figures. Remember to keep the numbers in the correct columns. The first one has been done for you.

1. 2·4 + 3·1 + 14·3 = *
2. 21·2 + 6·3 + 11·3 = *
3. 12·5 + 4·0 + 10·2 = *
4. 5·3 + 11·8 + 12·2 = *
5. 0·8 + 13·6 + 22·4 = *
6. 10·9 + 8·3 + 0·9 = *

Subtraction

Exercise 7

1. 8·6
   − 3·4

2. 9·5
   − 3·4

3. 8·0
   − 1·5

4. 7·2
   − 2·6

5. 9·0
   − 4·5

6. 15·8
   − 4·8

7. 28·6
   − 17·4

8. 28·0
   − 14·3

9. 42·3
   − 20·5

10. 33·2
    − 12·8

11. 50·3
    − 24·3

12. 31·6
    − 26·0

Copy and complete this table about your answers above.

Question number	1	2	3	4	5	6	7	8	9	10	11	12
How many tens in your answer?	0	*	*	*	*	1	*	*	2	*	*	*
How many units in your answer?	5	*	6	*	*	*	*	3	*	*	*	*
How many tenths in your answer?	2	*	*	*	5	*	*	*	*	*	*	*

Multiplying decimals

This is a drawing of David Croakett. In 1976 he made the world record frog jump. He jumped almost 5·2 m.

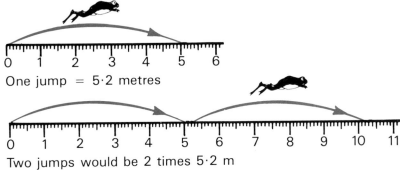

One jump = 5·2 metres

Two jumps would be 2 times 5·2 m

$$2 \text{ jumps} = 2 \times 5\cdot2 \text{ m} = \begin{array}{r} 5\cdot2 \text{ m} \times \\ 2 \\ \hline 10\cdot4 \text{ m} \\ \hline \end{array}$$

$$3 \text{ jumps} = 3 \times 5\cdot2 \text{ m} = \begin{array}{r} 5\cdot2 \text{ m} \times \\ 3 \\ \hline 15\cdot6 \text{ m} \\ \hline \end{array}$$

Exercise 8

Calculate the total length of these jumps. The first one is done for you.

$$\textbf{1. } 3 \text{ jumps each } 1\cdot3 \text{ m long} = \begin{array}{r} 1\cdot3 \text{ m} \times \\ 3 \\ \hline 3\cdot9 \text{ m} \\ \hline \end{array}$$

2. 2 jumps each 4·3 m long   **3.** 5 jumps each 4·1 m long

4. 4 jumps each 4·2 m long   **5.** 7 jumps each 5·1 m long

Exercise 9

Work out the total length of these jumps. Set out the questions like the examples above

1. 3·3 m x 2   **2.** 4·2 m x 3   **3.** 5·2 m x 4   **4.** 7·2 m x 4

(Remember to carry over in the following questions)

5. 4·6 m x 2   **6.** 5·6 m x 3   **7.** 4·5 m x 5   **8.** 7·6 m x 6

9. 10·4 m x 3   **10.** 12·4 m x 4   **11.** 6·7 m x 5   **12.** 11·9 m x 2

13. 21·5 m x 3   **14.** 23·7 m x 3   **15.** 16·8 m x 2   **16.** 31·6 m x 4

Division of decimals

The two parcels weigh 2·6 kg
1 parcel weighs: 2·6 kg ÷ 2

$$\begin{array}{r} 1\cdot 3 \text{ kg} \\ 2\overline{\smash{)}2\cdot 6} \end{array}$$

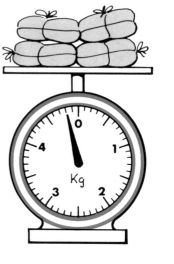

The four parcels weigh 4·8 kg
1 parcel weighs: 4·8 kg ÷ 4

$$\begin{array}{r} 1\cdot 2 \text{ kg} \\ 4\overline{\smash{)}4\cdot 8} \end{array}$$

Exercise 10

Find the weight of each parcel in these questions. The first one has been done for you.

1. 2 parcels weigh 6·4 kg
   1 parcel weighs $\begin{array}{r} 3\cdot 2 \text{ kg} \\ 2\overline{\smash{)}6\cdot 4} \end{array}$

2. 2 parcels weigh 8·8 kg

3. 4 parcels weigh 8·4 kg

4. 3 parcels weigh 9·6 kg

5. 3 parcels weigh 6·9 kg

Workcard 1	
1. 2)2·4	2. 2)4·2
3. 2)6·4	4. 2)4·6
5. 3)3·6	6. 3)6·6
7. 2)8·2	8. 2)6·0

Workcard 2	
1. 3)9·3	2. 3)3·9
3. 3)6·9	4. 2)6·8
5. 2)12·8	6. 3)12·6
7. 2)14·2	8. 2)7·2

Workcard 3	
1. 2)5·4	2. 3)7·2
3. 4)12·8	4. 4)13·6
5. 2)11·8	6. 4)22·4
7. 5)26·5	8. 6)14·4

Exercise 11

1. At the Chewo Sweet factory, Mary has to pack 19·2 kg of sweets into 6 boxes. How many kg will she put into each box?

2. At the Fizzpop Drinks factory, Tom has to put 22·5 litres of shandy into 5 jars. How many litres will he put into each jar?

3. At the Lux Lace factory, Robin has to cut 7·8 metres of ribbon into 6 equal lengths. How long will each piece be?

4. On this weighing scale are 4 boxes of fruit. How much does each box weigh?

5. Five workers in an office win £26·55 between them, in a raffle. How much do they each receive?

Workcard 4

1. 3)13·2   2. 4)13·2

3. 4)17·6   4. 3)16·8

5. 2)19·4   6. 2)17·8

7. 4)14·4   8. 3)17·1

Workcard 5

1. 5)16·5   2. 5)12·5

3. 6)14·4   4. 7)16·1

5. 5)21·5   6. 4)18·4

7. 4)25·2   8. 6)20·4

Workcard 6

1. 3)23·4   2. 3)25·5

3. 5)27·5   4. 4)30·4

5. 3)26·1   6. 5)32·5

7. 4)33·2   8. 6)31·8

Workcard 7

1. 2)126·4   2. 3)156·9

3. 4)172·8   4. 4)144·4

5. 3)141·6   6. 2)153·2

7. 4)253·2   8. 5)217·5

Workcard 8

1. 4)217·2   2. 3)232·8

3. 3)194·4   4. 4)225·6

5. 5)271·5   6. 4)262·4

7. 6)205·2   8. 5)281·5

Workcard 9

1. 4)21·72   2. 3)19·35

3. 4)26·16   4. 5)28·15

5. 3)25·71   6. 6)19·50

7. 5)32·65   8. 6)26·04

Review 1

A. Solids

1. Sort these solids into the correct rings. The first one is done for you.

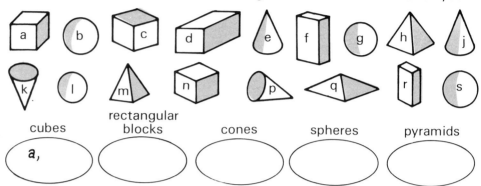

cubes

rectangular blocks

cones

spheres

pyramids

a,

2. What solid shapes will these nets produce when folded?

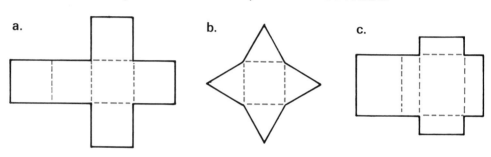

a.

b.

c.

B. Volume

1.

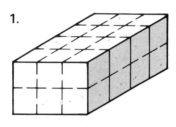

a. How many centimetre-cubes would it take to construct this shape?

b. What is the volume of this shape?

2.

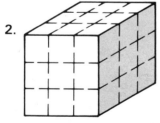

a. How many centimetre-cubes would it take to construct this shape?

b. What is the volume of this shape?

3. What are the volumes of these shapes?

a.

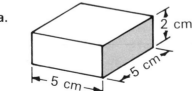

2 cm
5 cm
5 cm

b.

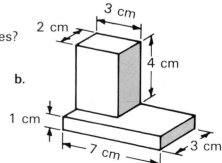

3 cm
2 cm
4 cm
1 cm
7 cm
3 cm

C. Angles in a circle

1. There are _____° in a circle.

2.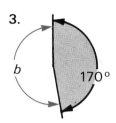

Angle $a = *$ °

3.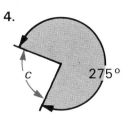

Angle $b = *$ °

4.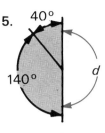

Angle $c = *$ °

5.

Angle $d = *$ °

D. Angles and parallel lines

By comparing the angles in each diagram below, say which drawings show a pair of parallel lines.

1.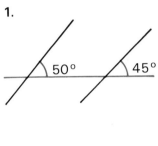

2.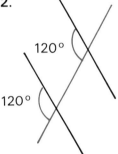

3.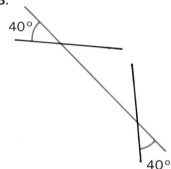

E. Fractions

1. From a group of 12 people, 6 cycle to work. What fraction cycle to work?

2. Roger saved £15. He spent £5 of this. What fraction did he spend?

3. If 9 pupils go home to lunch, 7 pupils have a packed lunch, and 5 pupils have lunch in a local café, what fraction have packed lunch?

4. Re-write these fractions in order of size, from biggest to smallest.

   a. $\frac{1}{5}$, $\frac{1}{2}$, $\frac{1}{4}$

   b. $\frac{1}{10}$, $\frac{1}{3}$, $\frac{2}{3}$

   c. $\frac{1}{4}$, $\frac{1}{2}$, $\frac{1}{10}$, $\frac{3}{4}$

F. Decimals

1.  If one bag of rice weighs 1·2 kg how much will 4 bags weigh?

2.  If one bag of flour weighs 1·5 kg, how much will 5 bags weigh?

3. If 5·2 kg of soil is shared between 2 flower pots, how much is put into each pot?

4. If a bucket can carry 4 kg of sand, how many buckets can be filled from 9·6 kg of sand?

G. Sets and sorting

1. a. Which of the two sets below is the subset?

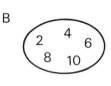

A 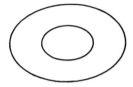 B

b. Re-draw the two sets like this.
Put the numbers on the diagram
in their correct position.

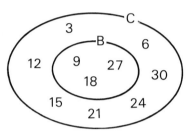

2. a. Make a list of the members of set C.
b. Make a list of the members of subset B.
c. Which members belong in set C but not
in set B?
d. The numbers in set C belong in the
_____ times table and the numbers in
set B belong to the _____ times table.

H. Minus numbers

1. Copy the number line below.

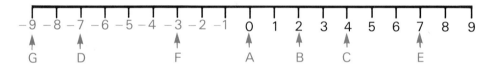

2. Write down the positions of the letters A to G.

3. Add arrows and letters to your number line showing: H at (-1)
and K at (-5)

4. Calculate or count the distance between these points on the number
line: **a.** A and B   **b.** A and E   **c.** B and E   **d.** A and F
e. F and D   **f.** F and G   **g.** B and F   **h.** D and C

5. Copy and complete these calculations. Use the number line to help.

a. $7 - 6 = *$ **b.** $6 - 7 = *$ **c.** $5 - 8 = *$

d. $(-2) + 4 = *$ **e.** $(-7) + 4 = *$ **f.** $(-9) + 9 = *$

g. $(-8) + 10 = *$ **h.** $(-2) - 2 = *$ **i.** $(-4) + * = 1$

J. Brackets

Answer these questions. Remember to do the calculations in brackets
first.

1. $(3 \times 4) + 6 = *$ **2.** $(5 \times 4) + 13 = *$ **3.** $12 + (6 \times 3) = *$

4. $(2 + 1) \times 5 = *$ **5.** $(2 + 8) \times 6 = *$ **6.** $4 \times (10 + 2) = *$

7. $(9 - 5) \times 4 = *$ **8.** $(14 - 3) \times 3 = *$ **9.** $28 - (3 \times 5) = *$

Exercise 1

Find the weight of each can. The first one is done for you.

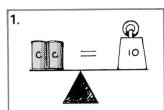

Use c for the weight
of each can
$2c = 10$
So $c = 5$
Each can weighs 5

2.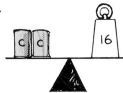

3.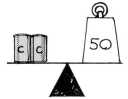

4.

5.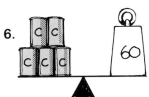

6.

7.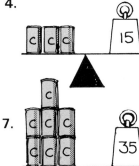

Exercise 2

Find the weight of each bottle. The first one is done for you.

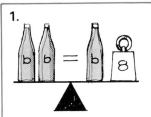

Use b for the weight
of each bottle
$2b = b + 8$
Take b away from each
side then $b = 8$

2.

3.

4.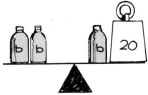

5. $2b = b + 6$
8. $3b = 2b + 10$

6. $2b = b + 15$
9. $3b = 2b + 12$

7. $2b = b + 25$
10. $3b = 2b + 19$

Exercise 3

Find the weight of each parcel. The first one is done for you.

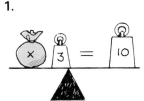

Now there are
numbers on both sides
of the balance
$x + 3 = 10$
So take 3 away from
each side
$x = 7$

2.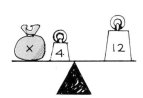

3.

4.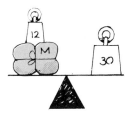

5.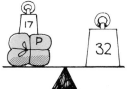

6.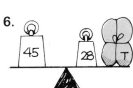

7.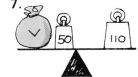

Exercise 4

Find the number of cans packed into the box in each question.
The first one is done for you. A symbol has been used to stand for the
number of cans in each box.

1.

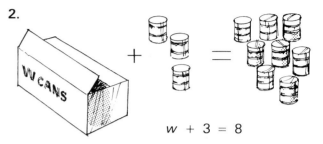

$$x + 1 = 7$$
(take one can away from each side)
$$x = 6$$

2.

$$w + 3 = 8$$

3.

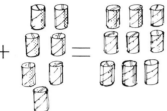

$$y + 7 = 9$$

4.

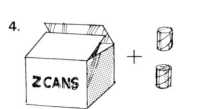

$$z + 2 = 4$$

Exercise 5

For each question find the number of matches in each box.
A symbol has been used to stand for the number of matches in each
box.

1.

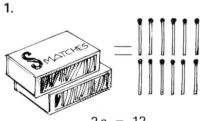

$$2s = 12$$

2.

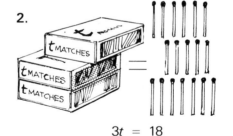

$$3t = 18$$

3.

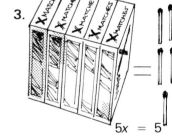

$$5x = 5$$

Exercise 6

For each question find the number of sweets in each tin.
A symbol has been used to stand for the number of sweets in each tin.

1.

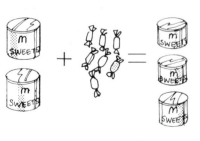

$$2m + 7 = 3m$$

2.

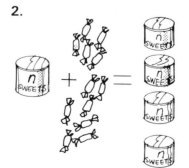

$$n + 12 = 4n$$

3.

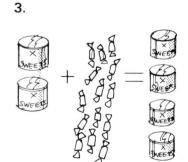

$$2x + 14 = 4x$$

There are cans and weights on both sides of this balance.
We want to find the weight of one can.

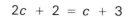

$$2c + 2 = c + 3$$

Take one can from each side of the balance.
The two sides still balance.

$$c + 2 = 3$$

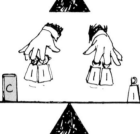

Now take two weights from each side of the balance.
The two sides still balance.

$$c = 1$$

One can weighs 1

Exercise 7

Calculate the weight of each can in these questions.

1.

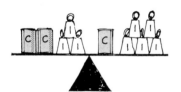

2.

3.

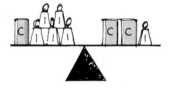

4.

5.

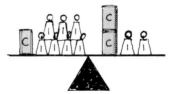

6.

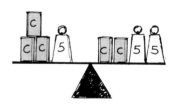

Exercise 8

Find the value of the letters in each question

1. $3c + 2 = 2c + 7$

2. $3p + 6 = 2p + 10$

3. $5t + 8 = 4t + 12$

4. $4f + 6 = 3f + 7$

5. $7a + 2 = 6a + 9$

6. $9v + 10 = 8v + 15$

7. $8r + 12 = 7r + 18$

8. $10q + 10 = 9q + 17$

9. $2e + 10 = 18 + e$

Jenny Strong is the wages clerk at the Clogworth Fashion factory.
She has to work out the wages of each of the workers.
Each worker gets a basic wage of £17 and a
bonus of £2 for each dress he or she makes.
So, if a worker makes 3 dresses in a week the bonus will be

$$3 \times £2 = £6.$$

Exercise 9

A worker made the following numbers of dresses.
Calculate the bonuses when these numbers of dresses are made.

1. 5 dresses   **2.** 10 dresses   **3.** 7 dresses   **4.** 12 dresses

5. 15 dresses   **6.** 20 dresses   **7.** 19 dresses   **8.** 25 dresses

Each time Jenny had to calculate a total wage
She would do it like this. It was too awkward

6 dresses:
6 dresses at £2 each is £12, and £17 basic
wage is a total of £29.

Exercise 10

Using the numbers of dresses in Exercise 9, work out the total wages
paid.

Jenny found an easier way to do her calculations.
She found a formula

The formula
$w = 2n + 17$

total wage

number of dresses

£17 basic wage

Using the formula $w = 2n + 17$ for 16 dresses
$w = (2 \times 16) + 17$
$w = 32 + 17$
$w = 49$
The total wage is £49

Exercise 11

Use the formula $w = 2n + 17$ to work out the wages for these workers.

1. Mavis made 9 dresses   **2.** Harry made 19 dresses   **3.** Tom made 11 dresses

4. Leo made 13 dresses   **5.** Keran made 16 dresses   **6.** Jan made 21 dresses

The workers are given a rise. They are paid £5 for each dress and a basic wage of £21 per week.
The new formula is $w = 5n + 21$

Exercise 12

Use the formula $w = 5n + 21$ to find the wages when:

1. $n = 2$ **2.** $n = 4$ **3.** $n = 6$ **4.** $n = 1$

5. $n = 3$ **6.** $n = 5$ **7.** $n = 7$ **8.** $n = 10$

9. $n = 12$ **10.** $n = 14$ **11.** $n = 20$ **12.** $n = 21$

The formula for wages is changed again. This time the workers are paid £6 for each dress and a basic wage of £18 per week.
The new formula is $w = 6n + 18$

Exercise 13

Use the formula $w = 6n + 18$ to find the wages when:

1. $n = 1$ **2.** $n = 3$ **3.** $n = 4$ **4.** $n = 2$

5. $n = 5$ **6.** $n = 6$ **7.** $n = 7$ **8.** $n = 8$

9. $n = 10$ **10.** $n = 9$ **11.** $n = 1\frac{1}{2}$ **12.** $n = 2\frac{1}{2}$

Exercise 14

1. Janice packs dresses into boxes. Her basic wage is £45 a week, and her bonus is £3 for each box she packs. Write the formula to calculate her total wage.

In your answer use
w for total wage
n for number of boxes packed

2. Bernie drives the dresses to shops. His basic wage is £69 a week, and his bonus is £5 for every delivery he makes. Write the formula to calculate his total wage.

In your answer use
w for total wage
n for number of deliveries

3. Margot sells dresses in a shop. Her basic wage is £58 a week, and her bonus is £1·50 for every dress she sells. Write the formula to calculate her total wage.

In your answer use
w for total wage
n for number of dresses sold.

Estimation and Rounding Off

An estimate is an approximate or rough calculation.
At a fête you could be asked to guess the weight of a cake or how many peas there are in a jar.

You cannot count the peas or weigh the cake so all you can do is give an estimate.

Exercise 1

Give an estimated answer to these problems.

1.

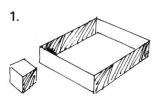

How many cubes do you think would fit into this tray?

2.

How many people do you think are on this big dipper?

3.

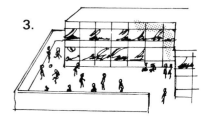

How many children do you think are in this playground?

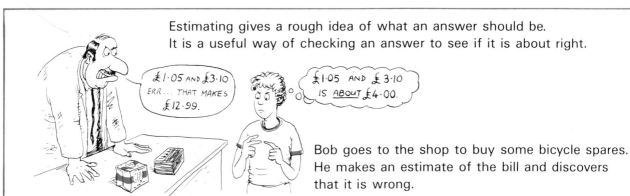

Estimating gives a rough idea of what an answer should be.
It is a useful way of checking an answer to see if it is about right.

Bob goes to the shop to buy some bicycle spares. He makes an estimate of the bill and discovers that it is wrong.

Exercise 2

Give an estimated answer to each of these questions.

1. How tall are you?
2. What is your weight?
3. How long does it take you to travel to school?
4. How far is your home from school?
5. How many pupils are there in your school?
6. Estimate how many teachers there are in your school.
7. How many seats are there on a double-decker bus?
8. How many hours per week do you spend asleep?
9. How old is your home?

It is useful to have a rough idea of your answer before you start a calculation. In this way you can avoid silly answers.

Exercise 3

Below are shown a number of bills. Decide which estimated total is the most accurate.

1.

EXPRESS CAFE

£1·00
£0·95
£2·26

TOTAL *·**

a. 'I think the total is about £3·00'

b. 'I think the total is about £6·00'

c. 'I think the total is about £4·00'

2.

RITZ CINEMA £3·90 STALLS
RITZ CINEMA £3·90 STALLS
RITZ CINEMA £3·90 STALLS

a. 'I think the total is about £12·00'

b. 'I think the total is about £9·00'

c. 'I think the total is about £10.00'

3.

Harry's Hardware Ltd

Pck of Nails   55p
Pck of Screws   £1·05p
Pck of Screws   £4·99p
Trowel   £1·10p
Wall Plugs   £
Total   *·**

a. 'I think the total is about £6·00'

b. 'I think the total is about £7·50'

c. 'I think the total is about £8·50'

4.

RAPID TRAVEL COACHES

2 Adults   £19-80
2 Children   £ 8·90
V.A.T   £ 4·31
      Total   **·**

a. 'I think the total is about £33·00'

b. 'I think the total is about £28·00'

c. 'I think the total is about £39·00'

5.

RIPLEYS SPARES DEPT.

Oil   £ 8·75
Filter   £ 7·99
Clamps   £15·25
V.A.T   £ 4·79
Total   **·**

a. 'I think the total is about £412·50'

b. 'I think the total is about £32·00'

c. 'I think the total is about £36·50'

Exercise 4

Estimate which answer is the most accurate.

1. $6·5 \times 2 =$ | 130. | 65. | 13. |

2. $18 + 81 =$ | 88. | 99. | -63. |

3. $397 - 99 =$ | 357. | 298. | 29.8 |

4. $500 \div 4 =$ | 125. | 2000. | 100. |

5. $224 + 8·6 =$ | 215.4 | 310. | 232.6 |

You can make an estimate of the number shown on the dial.
The dial on the right shows about 2·4

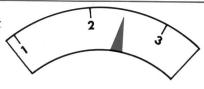

Exercise 5

Estimate the number shown on each of the dials below.

1.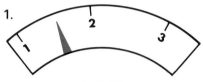

2.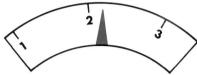

3.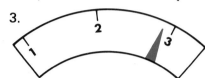

4.

5.

6.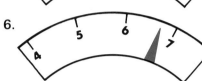

7.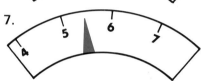

8.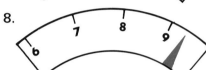

You can make an estimate of the amount in a container.
The container on the right is about $\frac{1}{2}$ full.

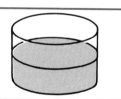

Exercise 6

Estimate the amount in each of the containers below. Choose your answer from the fractions shown in brackets.

1.

This fish tank is about * full.
$(\frac{1}{2}, \frac{1}{10}, \frac{1}{3})$

2.

This jug is about * full.
$(\frac{1}{4}, \frac{2}{3}, \frac{1}{3})$

3.

This mug is about * full.
$(\frac{3}{4}, \frac{1}{5}, \frac{1}{2})$

4.

This beaker is about * full.
$(\frac{1}{4}, \frac{1}{3}, \frac{1}{10})$

5.

This bottle is about * full.
$(\frac{2}{5}, \frac{2}{3}, \frac{3}{4})$

6.

This jar is about * full.
$(\frac{4}{5}, \frac{1}{4}, \frac{1}{2})$

There are times when it is useful to give an answer that has been rounded-off.

Janet asked two people for the time. One answer was very accurate, the other answer was rounded-off.

WHAT IS THE TIME, PLEASE?

IT IS FOUR MINUTES AND 27 SECONDS PAST 6

IT IS ABOUT FIVE PAST SIX

Exercise 7

Decide which answer, A or B has been rounded-off.

1. Mr Brown asks the Green twins their age.

HOW OLD ARE YOU?

Ⓐ I AM 15 YEARS OLD

Ⓑ I AM 15 YEARS 1 MONTH 4 DAYS AND 7 HOURS OLD.

2. Fred asks two men at the station, when the train will arrive.

WHEN WILL THE TRAIN ARRIVE?

Ⓐ IN PRECISELY 4 MINUTES AND 54 SECONDS

Ⓑ IT SHOULD BE HERE IN ABOUT 5 MINUTES.

3. Jenny asks two people how far it is to the bank.

HOW FAR AWAY IS THE BANK?

Ⓐ IT IS TWO KILOMETRES AND 625 METRES.

Ⓑ IT'S ABOUT TWO AND A HALF KILOMETRES FROM HERE.

Exercise 8

Measure these lines and round-off your answer to the nearest centimetre.

1. _____   2. _____

3. _____   4. _____

5. _____   6. _____

7. _____   8. _____

Exercise 9

Round-off these times to the nearest hour, half hour or quarter hour. The first one is done for you.

1.

It is about a quarter to 3.

2.   **3.**   **4.**   **5.**   **6.**

7.   **8.**   **9.**

Look at the number line. 27 is nearer to 30 than 20. Rounded-off to the nearest ten 27 becomes 30. Round-off these numbers to the nearest ten.

Exercise 10

1. 11 is nearer to * (10 or 20)   **2.** 19 is nearer to * (10 or 20)

3. 17 is nearer to * (10 or 20)   **4.** 22 is nearer to * (20 or 30)

5. 33 is nearer to * (30 or 40)   **6.** 47 is nearer to * (40 or 50)

Exercise 11

Numbers ending in 5 are always rounded up. Round-off these numbers.

1. 15 → 20	**2.** 22	**3.** 31	**4.** 58
5. 75	**6.** 89	**7.** 56	**8.** 71
9. 5	**10.** 25	**11.** 68	**12.** 7
13. 85	**14.** 212	**15.** 155	**16.** 173

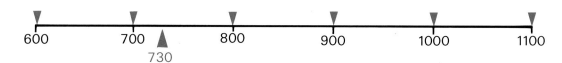

730 is nearer to 700 than 800. Rounded-off to the nearest hundred, 730 becomes 700. Round-off these numbers to the nearest hundred.

Exercise 12

1. 860 is nearer to * (800 or 900)   **2.** 620 is nearer to * (600 or 700)

3. 980 is nearer to * (900 or 1000)   **4.** 1190 is nearer to * (1100 or 1200)

5. 465 is nearer to * (400 or 500)   **6.** 1340 is nearer to * (1300 or 1400)

Exercise 13

Round-off these populations to the nearest hundred.

Place	Oban	Beddgelert	Studland	Usk	Holkham	Creetown
Population	8134	671	620	1890	272	785
Rounded-off	8100	*	*	*	*	*

Exercise 14

Round-off these amounts to give convenient answers.

1. Our school goalie let in 302 goals this year.

2. My uncle Bob scored 3796 runs this season.

3. I am 4988 days old.

4. Mount Everest is 8848 metres high.

3 + 4 + 6 = ⟦ ? ⟧

To solve this problem with a calculator
take the following steps:

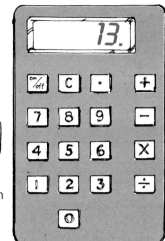

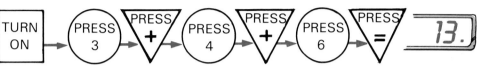

TURN ON → PRESS 3 → PRESS + → PRESS 4 → PRESS + → PRESS 6 → PRESS = → 13.

Exercise 1   Copy these flow charts and lay out each problem
within its flow chart.

1. 4 x 3 =

TURN ON → ◯ → ▽ → ◯ → ▽ → ⟦ANSWER⟧

2. 9 + 6 − 3 =

▢ → ◯ → ▽ → ▽ → ◯ → ▽ = ⟦ANSWER⟧

3. 27 + 52 =

▢ → ◯ → ◯ → ▽ → ◯ → ◯ → ▽ → ⟦ANSWER⟧

4. 64 ÷ 2 =

▢ → ◯ → ◯ → ▽ → ◯ → ▽ → ⟦ANSWER⟧

5. 9 + 7 − 5 =

▢ → ◯ → ▽ → ◯ → ▽ → ◯ → ▽ → ⟦ANSWER⟧

Exercise 2   How many of these problems can you do in one minute?
Only write down your answer.

> If you make
> an error press
> C to cancel
> your last entry

1. 4 + 8 = *   **2.** 9 − 5 = *   **3.** 6 x 3 = *   **4.** 8 ÷ 4 = *

5. 11 − 2 = *   **6.** 8 x 4 = *   **7.** 12 ÷ 2 = *   **8.** 18 + 7 = *

9. 5 x 7 = *   **10.** 20 ÷ 5 = *   **11.** 17 + 23 = *   **12.** 31 − 24 = *

13. 36 ÷ 6 = *   **14.** 19 + 19 = *   **15.** 43 − 37 = *   **16.** 22 x 2 = *

Calculators and decimals

Calculators can be used to work out problems with decimal fractions.
These are the steps to take to solve this problem on a calculator.

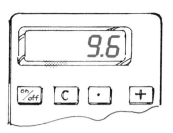

$3 \times 3.2 =$

Exercise 3

Copy the questions into your book and answer them.

1. $4 \times 4.2 = *$ **2.** $3 \times 3.4 = *$ **3.** $5 \times 4.6 = *$ **4.** $7 \times 2.4 = *$

5. $3 \times 5.6 = *$ **6.** $5 \times 8.3 = *$ **7.** $9.6 \times 2 = *$ **8.** $8.4 \times 3 = *$

9. $7.7 \times 5 = *$ **10.** $10.3 \times 3 = *$ **11.** $12 \times 3.1 = *$ **12.** $10.5 \times 4 = *$

13. $12.5 \times 2 = *$ **14.** $4 \times 11.5 = *$ **15.** $14 \times 2.2 = *$ **16.** $5.7 \times 12 = *$

Calculators and money

The total cost of these three items
is £4·40. Check this total on your
calculator.
Your answer will look like the
one on the drawing opposite: 4.4

Calculators do not show pound signs (£).
When a decimal answer ends in 0, calculators do not show the 0. So $4.4 = £4.40$

Exercise 4

Write out the sums of money shown on these displays. These amounts
are all in pounds.

1.    **2.**    **3.**

4.    **5.** 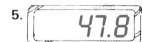   **6.**

7. 0.8   **8.** 0.3   **9.** 16.6

10. 18.   **11.** 52.   **12.**

This display shows £7   This display shows 70p   This display shows 7p
                        (7 x 10p)   (7 x 1p)

Notice how the position of the decimal point affects the value shown on the display.

Exercise 5

Answer the questions about these displays.

1.
a. How many £'s are shown?
b. How many 10p's are shown?
c. How many pennies are shown?

2.
a. How many £'s are shown?
b. How many 10p's are shown?
c. How many pennies are shown?

3.
a. How many £'s are shown?
b. How many 10p's are shown?
c. How many pennies are shown?

4.
a. How many £'s are shown?
b. How many 10p's are shown?
c. How many pennies are shown?

Exercise 6

Add these amounts using a calculator.

1. £0·47 + £0·26 =   2. £0·33 + £0·13 =   3. £0·40 + £0·04 =

4. £0·62 + £0·37 =   5. £0·09 + £0·01 =   6. £0·96 + £0·02 =

7. £0·31 + £0·09 =   8. £0·08 + £0·09 =   9. £0·66 + £0·06 =

10. £1·50 + £2·06 =   11. £0·06 + £1·24 =   12. £2·09 + £1·06 =

13. £3·08 + £2·72 =   14. £0·50 + £0·50 =   15. £1·95 + £1·05 =

Exercise 7

comb 49p   shampoo 99p   hairgrips £1·01   soap 62p   perfume £6·05   toothpaste 70p

Using a calculator, work out the bills for these purchases.

1. soap, shampoo, hairgrips = *   2. toothpaste, comb, hairgrips = *

3. perfume, comb = *   4. hairgrips, perfume = *

5. shampoo, hairgrips = *   6. hairgrips, soap, perfume = *

7. comb, soap, perfume = *   8. toothpaste, perfume, shampoo = *

9. comb, hairgrips, soap = *   10. shampoo, hairgrips, perfume = *

11. How much change from £1·00 would you get if you bought one bar of soap?

12. How much change from £1·00 would you get if you bought one comb?

> Which of these two calculations is the quickest;
>   **a.** 4·65 + 4·65 + 4·65 + 4·65 + 4·65 + 4·65
> or **b.** 4·65 x 6
> Try them both on your calculator.
> Do both methods give the same answer?

Exercise 8   Copy and complete the calculations below. Do not work out the answers.

1. 4·51 + 4·51 + 4·51 + 4·51 = 4·51 x

2. 17·2 + 17·2 + 17·2 + 17·2 = 4 x

3. 9·75 + 9·75 + 9·75 + 9·75 + 9·75 + 9·75 + 9·75 = ⬭ x ⬭

4. 0·38 + 0·38 + 0·38 + 0·38 + 0·38 + 0·38 = ⬭ x ⬭

5. 10·2 + 10·2 + 10·2 + 10·2 + 10·2 = ⬭ x ⬭

£1·95

£0·45   £10·69

£2·20

Exercise 9   Work out the cost of these bills:

1. two records   **2.** four magazines

3. five tape cassettes   **4.** three walkman machines

5. seven records   **6.** nine hats

7. six magazines   **8.** five walkman machines

 Add up the bill for these items.
First find the cost of the two records and make a note of the amount.
Then find the cost of the three magazines. Finally add the two
amounts together and find the total.
This can be made more easy if your calculator has a memory button.
Your teacher will show you how to use the memory.

Exercise 10   Work out these bills, using the process shown above.

1. two records and three magazines   **2.** four hats and five cassettes

3. two walkmans and six cassettes   **4.** nine magazines and four records

5. seven hats and five magazines   **6.** eight cassettes and one walkman

$5 + 9 \times 4$

There are two ways of doing this problem. They give different answers.

| $(5 + 9) \times 4 = 56$ | or | $(9 \times 4) + 5 = 41$ |

$5 + 9 = 14$

$14 \times 4 = 56$

$9 \times 4 = 36$

$36 + 5 = 41$

Do these problems on your calculator like this

1st way:

2nd way:

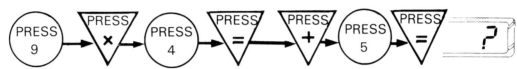

Always do the sum in brackets first, press the ⊟ button then do the next part of the problem.

Exercise 11

Use your calculator to answer these problems.

1. $(15 + 6) \times 5 = *$ **2.** $(20 - 7) \times 6 = *$ **3.** $(4 \times 5) + 16 = *$ **4.** $(28 \div 14) + 19 = *$

5. $3 \times (16 + 9) = *$ **6.** $20 + (36 \div 9) = *$ **7.** $15 + (220 \div 4) = *$ **8.** $(125 \div 5) - 25 = *$

To convert fractions to decimals we use the ⊟ button

$\frac{1}{5}$ means $1 \div 5$ PRESS 1 → PRESS ÷ → PRESS 5 → PRESS =   *0.2*

Exercise 12

Convert these fractions to decimals

1. $\frac{1}{2}$ **2.** $\frac{1}{10}$ **3.** $\frac{1}{4}$ **4.** $\frac{3}{4}$ **5.** $\frac{1}{8}$ **6.** $\frac{1}{3}$ **7.** $\frac{2}{3}$

Look at your answer to questions 6 & 7 above. The answers have used all the digits on the display *0.333333*   *0.666666*
Only the first two digits after the decimal point in a number like this are useful, so we round-off the numbers.

     $0.333333 \ldots$ is written 0.33 but $0.66666 \ldots$ is written 0.67
     because $0.66666 \ldots$ is nearer 0.67 than 0.660.

Exercise 13

Convert these fractions to decimals. Round-off the answers so that they have two digits after the decimal point.

1. $\frac{1}{3}$ **2.** $\frac{1}{6}$ **3.** $\frac{5}{6}$ **4.** $\frac{2}{3}$ **5.** $\frac{1}{8}$ **6.** $\frac{1}{7}$ **7.** $\frac{4}{7}$

8. $\frac{4}{9}$ **9.** $\frac{7}{8}$ **10.** $\frac{7}{9}$ **11.** $\frac{1}{11}$ **12.** $\frac{4}{11}$ **13.** $\frac{1}{12}$ **14.** $\frac{7}{12}$

Checking your answers

It is easy to make a mistake when you are working with a calculator especially if you are working quickly.

Try to make an estimate of your answer so that you will have a rough idea of whether your answer is correct

Like this:

The area of the rectangle is
10·5 cm x 5 cm

Estimate: 10·5 is between 10 and 11 so your answer should be between 50 cm² and 55 cm²

Exercise 14

For each question below make an estimate of the answer before you do the calculation

1.

10·5 m

There are 3 floors in the block
How high is each floor?

2.

Aylesbury

London 64 km

135 km

Bournemouth

How far is it from Aylesbury to Bournemouth?

3. Oxford

London 286 km

129 km

Brighton

How far is it from London to Oxford?

4.

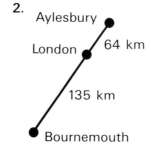

3 m

The area of the carpet is 18·6 m²
How long is the carpet?

5.

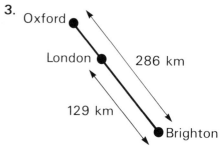

How many matches are there in 12 boxes?

6.

How much would 7 bags of flour weigh?

7.

The machine makes 80 sausages per minute. How many sausages will it make in 15 minutes?

8. If Joy earns £99·52 a week, how much does she earn in one year (52 weeks = 1 year)?

9. In an election 4500 people voted for 2 candidates. Green received 1967 votes. How many votes did Cartwright receive?

It is important that you should be able to check your answers.
The total of this bill is £6·74.
The customer paid £10·00 and received
£3·26 change.
To check whether the change is
correct you could make several
checks:

add the total and the change, they
should come to £10·00

take the change away from the £10·00, it should come to the total.

Exercise 15   Check the change in these bills. Rewrite the incorrect bills and show the
correct change

1.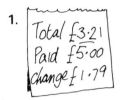

2.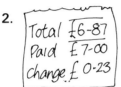

3.

4.

5.

6.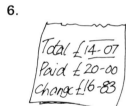

7.

8.

9.

10.

Exercise 16   Use your calculator to check the statements written in blue. Correct
sentences which are wrong.

1.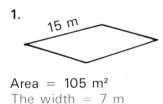

Area = 105 m²
The width = 7 m

2.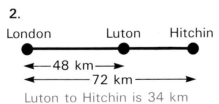

Luton to Hitchin is 34 km

3.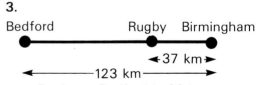

Rugby to Bedford is 86 km

4.

The load on the truck is 88 kg
Each box weighs 26 kg

5.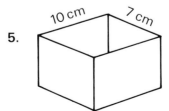

The volume of the box
is 210 cm³
The height of the box is 3 cm

6.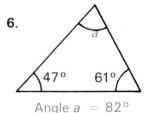

Angle a = 82°

'STAR NUMBERS'

Use your calculator to work out your Star Number.

1 With hard work you could become famous

2 You will marry young and have a large family

3 You tend to be lazy, but you will overcome this and be sucessful'

Example

1. Write out your date of birth in figures ➤ 26·4·78
2. Add the figures together ➤ 26 + 4 + 78 = 108
3. Count the letters in your name ➤ Graham = 6 letters
4. Add these two totals together ➤ 108 + 6 = 114
5. Multiply this new total by 3 ➤ 114 x 3 = 342
6. Add the figures in this total together ➤ 3 + 4 + 2 = 9
7. If your answer is more than 9, add the figures together once again.

4 Good luck is just around the corner

5 Time to make up your mind about an important matter

6 Someone nearby admires you greatly

7 Recent bad luck will end and turn out in your favour

8 You will enjoy your job and live to be very old

9 Some good news is coming your way

If you key a string of numbers into a calculator and then look at them upside-down, some of the numbers look like letters of the alphabet.

O   B   L   S   h   E   I

Key 0·7734 into a calculator and look at it up-side down.

You should see this:— hELL·O

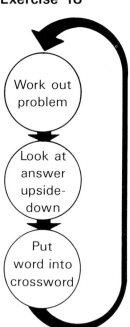

Work out problem

Look at answer upside-down

Put word into crossword

Copy the crossword grid below. Work out each problem. Look at the answer upside-down. Enter the word into the crossword.

CLUES

Across

1. (9900 + 9009) x 2 : Good book
4. 4000 − 249 : ____ of Man
7. 1547 x 5 : Trade for money
8. 2000 − 920 + 2000 : Musical instrument

Down

1. 13844·5 x 4 : Good wishes
2. (500 x 14) + 108 : Cook in water
3. 1832·5 ÷ 2·5: : Snake-like fish
5. 24225 ÷ 3 : Untidy person
6. 30 + 3400 + 143 : Not me, someone ____

Jon Minnoch weighed probably more than 635 kg (100 st.)! When he became unwell, rescuers had to knock down the front of his home to get him to hospital. They didn't have a stretcher big enough to carry him so they used planks.

Lucin Zarate was the lightest person on record. At 17 years of age she weighed just over 2 kg ($4\frac{1}{2}$ lbs) At 20 years of age she weighed 5·9 kg (13 lbs).

We eat food to give us energy. We measure this energy in units called Calories. If we eat more Calories than we need, we put on weight. If we eat less Calories than we need, we lose weight. An adult needs between 2000 and 3000 Calories a day.

Foods and their Calorie content

Food	Cals	Food	Cals	Food	Cals
Apple	40 Cals	Cola (can)	130 Cals	Peas (1 oz)	15 Cals
Bacon (one slice)	100 Cals	Crisps (bag)	150 Cals	Pie	1240 Cals
Baked beans (1 oz)	20 Cals	Cocoa (mug)	115 Cals	Pizza (small)	305 Cals
Banana	65 Cals	Egg (boiled)	80 Cals	Potato (boiled)	23 Cals
Beef and kidney pie	525 Cals	Egg (fried)	100 Cals	Potato chips (per	
Bread and butter	40 Cals	Fish finger	55 Cals	portion)	240 Cals
Burger (4 oz)	430 Cals	Fried fish	330 Cals	Sausage (big, pork)	165 Cals
Cauliflower (1 oz)	3 Cals	Lemonade (can)	80 Cals	Sausage roll	290 Cals
Chocolate (1 bar)	420 Cals	Milk ($\frac{1}{2}$ pint)	180 Cals	Shepherd's pie	500 Cals
Coffee with milk	30 Cals	Orangeade (can)	110 Cals	Sugar (teaspoon)	30 Cals
				Tea with milk	30 Cals

Exercise 1

1. Find the number of Calories in these four meals.

breakfast

lunch snack

dinner

supper

2. Make up four meals like the ones above. Use the chart to add up the total number of Calories

3. Make up a vegetarian meal. What is the total number of Calories in it?

When we burn up energy, we use up Calories. All activity uses up Calories. Some activities use up Calories quicker than others.

This list shows how many Calories are used per minute (approximately) doing these activities.

Dancing	5	Lacrosse	6	Basketball	6	Writing	2
Soccer	6	Running	13	Golf	4	Cycling	6
Hockey	6	Tennis	6	Washing up	5	Canoeing	5
Badminton	4	Jogging	8	Walking	5	Making beds	7
Swimming	7	Cricket	3	Judo	6	Sleeping	1

Exercise 2

1. How many Calories would be used, writing for

   a. 8 minutes   b. 10 minutes   c. 15 minutes

2. How many Calories would be used jogging for

   a. 5 minutes   b. 10 minutes   c. 20 minutes

3. How many Calories would be used playing tennis for

   a. 5 minutes   b. 10 minutes   c. 15 minutes

Exercise 3

1. Joe cycles for 10 minutes, then makes his bed in 3 minutes, and then sleeps for 20 minutes. How many Calories has he used?

2. Jenny plays hockey for 20 minutes, goes running for 4 minutes and swims for 10 minutes. How many Calories has she used up?

3. Ronnie plays basketball for 10 minutes, jogs for 5 minutes and then plays 10 more minutes of basketball. How many Calories has he used?

Exercise 4

Use the table at the top of the page to decide which activity uses up the most Calories.

1. running for 5 minutes   or dancing for 20 minutes?
2. washing up for 15 minutes   or playing badminton for 10 minutes?
3. playing judo for 20 minutes or canoeing for 25 minutes?
4. sleeping for 1 hour   or playing cricket for fifteen minutes?

Exercise 5

1.

How many Calories does a basketball player use up in the period shown by the clocks?

2.

How many Calories would each of two judo fighters use in the period shown by the clocks?

3.

How many Calories would Fred use sleeping in the period shown on the clocks?

Exercise 6

1. How much do two boxes of mushrooms cost?

2. How much do two bags of celery cost?

3. How much do 3 lbs of turnips cost?

4. How much do 2 lbs of white potatoes cost?

5. How much do 2 lbs of parsnips cost?

6. How much do these bills add up to?
   - **a.** 1 lb of new potatoes
     1 lb of tomatoes
     1 lb of turnips
   - **b.** Box of mushrooms
     Bag of celery
     1 lb of carrots
   - **c.** $\frac{1}{2}$ lb of peppers
     1 lb of sprouts
     $\frac{1}{2}$ lb of carrots

Exercise 7

1.

How many boxes of Chilli Chips would you have to buy to get 120 g?

2.

If you get 6 eggs in a box, how many boxes would you buy to get 42 eggs?

3.

How many biscuits would you get in 5 packets?

4.

There are 10 cans of beans on this shelf. How many cans would be on a full shelf?

5.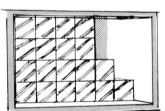

There are 19 boxes on this shelf. How many boxes would be on a full shelf?

6.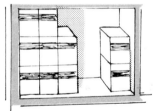

There are 10 packs on this shelf. How many packs would be on a full shelf?

Exercise 8

Which buy do you think is the best value, A or B?

1.

A   B

2.

A   B

3.

A   B

Cookbook

Here are the ingredients Rita used to make herself a mushroom omelette

2 eggs pinch of salt

$\frac{1}{2}$ a cup of milk 4 mushrooms

Exercise 9   Write down the ingredients needed for Rita to make omelettes for **a.** 3 people **b.** 4 people.

Joe makes some fruit salad for himself. Here are the ingredients that he used

3 grapes $\frac{1}{2}$ of a banana

2 cherries $\frac{1}{2}$ of an apple

1 peach 1 teaspoon of sherry

Exercise 10   Write down the ingredients that Joe would need to make enough for **a.** 4 people **b.** 6 people.

These are the ingredients needed to make a chicken curry for 2 people.

1 lb of chicken   4 tablespoons of curry powder
2 potatoes   3 tomatoes
1 tub of yoghurt   1 pint of chicken stock
5 small onions   6 cloves of garlic.   3 cups of rice.

Exercise 11   Copy and complete the table below

Ingredients	For 8 people	For 3 people
chicken	4 lbs	$1\frac{1}{2}$ lbs
spoons of curry powder	*	*
potatoes	*	*
tubs of yoghurt	*	*
pints of chicken stock	*	*
small onions	20	*
cloves of garlic	*	*
cups of rice	*	$4\frac{1}{2}$
tomatoes	*	*

Ratios

To make pastry for pies you need to mix
one part of lard with two equal parts
of flour.
We say the ratio of lard to flour is 1 to 2.
This means that if you used 50 g of lard you
would need 50 g x 2 = 100 g of flour.

Exercise 12

Using the ratio of 1 to 2 of lard to flour, complete the table below.

Parts of lard	Parts of flour
1	2
2	*
4	*
5	*
7	*
3	*
9	*
*	16
*	20
*	80
35	*
42	*

To make Butter Cream you mix 2 parts of butter with 3 parts of sugar.
This is a ratio of 2 to 3, written as 2 : 3.

Exercise 13

Using the ratio of butter to sugar as 2 : 3, copy and complete the following problems. The first one is done for you.

1. 4 parts butter : **6** parts sugar
2. 6 parts butter : * parts sugar
3. 8 parts butter : * parts sugar
4. 10 parts butter : * parts sugar
5. 20 parts butter : * parts sugar
6. 40 parts butter : * parts sugar

Exercise 14

Complete the problems. They are all in a ratio of 3 : 5. The first one is
done for you.

1. 3 : 5 = 6 : **10**
2. 3 : 5 = 9 : *
3. 3 : 5 = 18 : *
4. 3 : 5 = 12 : *
5. 3 : 5 = 15 : *
6. 3 : 5 = 30 : *

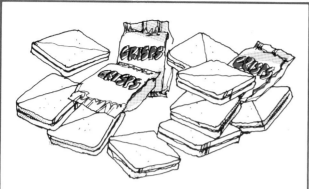

Here are 3 packets of crisps
and 9 sandwiches.
The ratio of crisps to sandwiches is 3 : 9

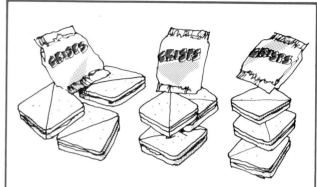

Notice that there is 1 packet of crisps
for every 3 sandwiches.
The ratio of crisps to sandwiches is 1 : 3

Exercise 15   Copy and complete the sentences below

1.
    a. The ratio of crisps to biscuits is 2 : *
    b. The ratio of crisps to biscuits is 1 : *
    c. The ratio of biscuits to crisps is 3 : *

2.
    a. The ratio of drinks to cakes is 3 : *
    b. The ratio of drinks to cakes is 1 : *
    c. The ratio of cakes to drinks is * : 1

3.
    a. The ratio of sandwiches to apples is * : 3
    b. The ratio of sandwiches to apples is 4 : *
    c. The ratio of apples to sandwiches is 1 : *

Exercise 16   Measure the lines a and b in each question. Write down the ratio of the
length of line a to the length of line b in the simplest form.

1. —— a   ———— b
2. —— a   ———— b
3. ———— a   —— b
4. ———— a   —— b
5. ————— a   — b
6. ———— a   ———— b

Review 2

A. Shape

 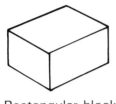  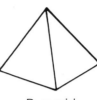 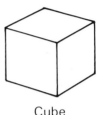

Cylinder   Rectangular block   Cone   Sphere   Pyramid   Cube

1. Each of these solids have been put into a sack. By reading the clues, say which solid is in which sack.

2. Which solids are used to construct each of these objects?

a. 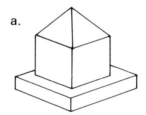   b. 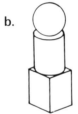   c.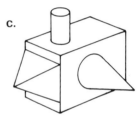

B. Volume

What is the volume of these rectangular blocks?

1.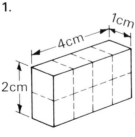

Volume = * cm³

2.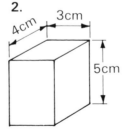

Volume = * cm³

3.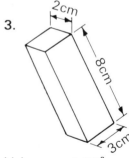

Volume = * cm³

What is the length of the missing measurement for each of these rectangular blocks?

4.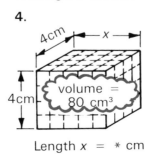

Length x = * cm

5.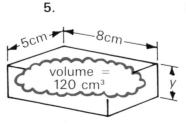

Length y = * cm

6.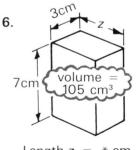

Length z = * cm

C. Ratio

What is the ratio of leaves to branches in each diagram below?

1. a.  **b.**  **c.**   **d.**

leaves to branches
* : 1

2. Which of these drawings show the ratio of eggs to slices of bacon as 2 : 3?

a.   **b.**   **c.**  **d.**

3. If you need three eggs to make 1 omelette how many eggs do you need to make 5 omelettes?

4. This recipe makes enough bread for 2 loaves. Re-write the recipe for 1 loaf.
2 kg of flour, 4 cups of water, 1 teaspoon of salt, 60 g of yeast, 30 g of bran

D. Fractions

1. Louise had £9 and she spent $\frac{1}{3}$ of it. How much did she spend?

2. Find half of these amounts:

   a. 18   **b.** £12   **c.** 24p   **d.** 20 cm   **e.** £30   **f.** 42 kg

3. Find a third of these amounts:

   a. 9   **b.** 24 kg   **c.** 18p   **d.** 36   **e.** £15   **f.** 30 cm

4. What is $\frac{2}{3}$ of 9?   **5.** What is $\frac{2}{5}$ of 20?   **6.** What is $\frac{3}{4}$ of 16?

7.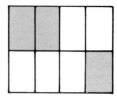

What fraction is coloured?

8.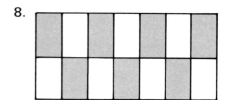

What fraction is coloured?

Probability and averages

Ken wants to go to the cinema, Ruth wants to go to the fun-fair.
To decide where to go they 'flip' a coin.
The result can be either 'heads' or 'tails'. Ken knows that his chances of winning are 'one chance out of two' or $\frac{1}{2}$.

Exercise 1

Ruth and Ken go to the fun-fair.
Ken plays 'Find the pea'.
There are 3 cups. A pea is hidden under one of them. Ken's chances of picking the correct cup are 'one chance out of three' or $\frac{1}{3}$.

1. If there were 5 cups Ken's chances would be:
one chance out of * or $\frac{1}{*}$

2. If there were 7 cups Ken's chances would be:
one chance out of * or $\frac{1}{*}$

3. If there were 4 cups Ken's chances would be:
one chance out of * or $\frac{1}{*}$

Exercise 2 What are Ruth's chances of winning in each of these games?

1. ROLLER DICE  Ruth's chances of picking the winning number are $\frac{1}{*}$	**2.** Wheel of Fortune  Ruth's chances of picking the winning number are $\frac{1}{*}$	**3.** Find the Ace  Ruth's chances of picking the ace are $\frac{1}{*}$	**4.** Lucky Numbers 19 7 22 6 11 32 10 37 17 2 29 3 8 16 4 28 5 18 14 33 25 Ruth's chances of picking the lucky number are $\frac{1}{*}$
5. Roulette  Ruth's chances of picking the winning number are $\frac{1}{*}$	**6.**  Ruth buys a raffle ticket. 150 tickets were sold. Ruth's chances of having the winning ticket are $\frac{1}{*}$.	**7.** Lucky Dip  One of these packages contains a present. Ruth's chances of picking the present are $\frac{1}{*}$.	**8.** SPIN & WIN The chances of the swinging arrow landing on the blue section are $\frac{1}{*}$

If the arrow is spinning on this wheel of fortune, the probability that it lands on the blue section is $\frac{1}{8}$ because there is only one blue section.

The probability of the arrow resting on a blue section on this wheel is $\frac{2}{8}$ because two out of the eight sections are blue. $\frac{2}{8}$ can also be written as $\frac{1}{4}$.

Exercise 3

Find the probability of the arrows resting on the blue sections. Write your answers as fractions

1. 2. 3. 4. 5. 6. 7.

8. Write as a fraction the probability of the arrow landing on the white sections of each drawing above.

Exercise 4

There are two 'black' cards in each hand below. What is the probability of picking a black card?

1. 2. 3. 4. 5. 6.

7. Which is the only hand that you will definitely pick a black card on the first go?

Exercise 5

Answer the questions below.

1. What are the chances of rolling a 'six'?

2. What is the probability of rolling a 'one'?

3. What is the probability of rolling a 'six' or a 'one'?

4. What are the chances of rolling a 'one', 'six', or 'three'?

5. What are the chances of rolling a 'two', 'six, 'five' or 'three'?

Exercise 6

1. There are 20 sweets in a bag, 5 of them are toffees.
   The chances of picking a toffee are ___

2. There are 100 light bulbs in a box. 2 of the bulbs are broken.
   The chances of picking out a broken bulb are ___

3. There are 80 marbles in a bag. 10 marbles are yellow.
   The chances of picking out a yellow marble are ___

4. There are 300 raffle tickets in a box. 100 are blue, 100 are red
   and 100 are green. The chances of picking out a green ticket are ___

5. In a pack of 52 playing cards there are 4 aces.
   The chances of picking out an ace are ___

Here are 4 statements. The first one is certain, two are possible and the
last one is impossible.

certainty	high probability	low probability	impossible
It is certain that you have a name.	It is highly likely that you will eat some food in the next 24 hours.	It is unlikely that you will find buried treasure tomorrow.	It is impossible that by flapping your hands you could fly.

Exercise 7

Copy out the four headings above: 'certainty', 'high probability', 'low
probability' and 'impossible'. Put each of the following statements under
the heading that you think is appropriate.

a. SOMEONE WILL WRITE YOU A LETTER SOON.

b. YOU COULD THROW A FEATHER FROM YOUR CLASSROOM AS FAR AS THE SUN.

c. YOU WERE BORN.

d. YOU WILL DRINK SOMETHING BEFORE MIDNIGHT TONIGHT.

e. A FLOWER WILL GROW OUT OF YOUR BEST FRIEND'S HEAD.

f. YOU WILL GET MARRIED.

g. YOU ARE LOOKING AT THIS BOOK. MATHSWISE 3

h. SOMEWHERE ON EARTH A PERSON IS TALKING. BLAH BLAH BLAB BLAB NATTER NATTER

i. YOU WILL WALK ON THE SUN TOMORROW.

j. SOMEONE WILL GIVE YOU £1000 THIS WEEK.

k. YOU WILL BECOME AN ELEPHANT TRAINER WHEN YOU LEAVE SCHOOL.

l. YOU WILL WATCH TELEVISION THIS EVENING.

Averages

Ali, Linda and Ben spend Saturday doing odd jobs in the local market. At the end of the day they agree to average out their earnings.

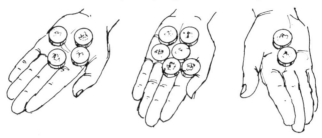

They have £12 between them. When they spread the money out evenly they each have £4. The average is £4.

Exercise 8

Spread each amount out evenly so that:

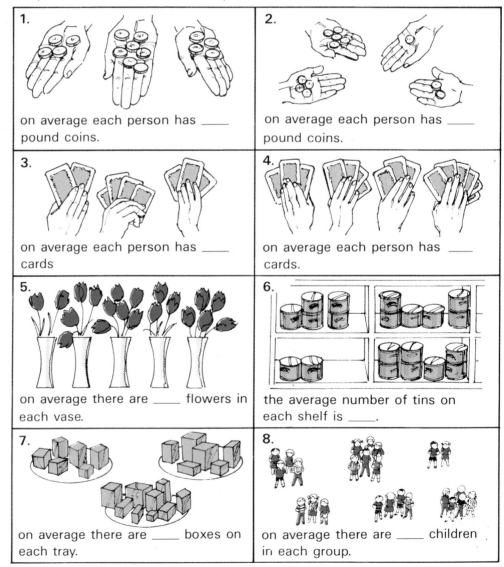

1. on average each person has ____ pound coins.

2. on average each person has ____ pound coins.

3. on average each person has ____ cards

4. on average each person has ____ cards.

5. on average there are ____ flowers in each vase.

6. the average number of tins on each shelf is ____.

7. on average there are ____ boxes on each tray.

8. on average there are ____ children in each group.

When we spread out the total number of items evenly we call this type of average the MEAN.

Farmer Jones has 4 hens. They lay 16 eggs between them.

The average for each hen is 16 ÷ 4 = 4 eggs.
So each hen laid an average of 4 eggs.

Exercise 9

1. If 3 hens laid 12 eggs, what was the mean laid by each hen?
2. If 5 hens laid 15 eggs, what was the mean laid by each hen?
3. If 4 hens laid 16 eggs, what was the mean laid by each hen?
4. If 6 hens laid 18 eggs, what was the mean laid by each hen?
5. If 4 hens laid 24 eggs, what was the mean laid by each hen?
6. If 7 hens laid 21 eggs, what was the mean laid by each hen?
7. If 8 hens laid 32 eggs, what was the mean laid by each hen?
8. If 3 hens laid 27 eggs, what was the mean laid by each hen?

Exercise 10

Copy and complete these sentences.

1.  The mean weight of the boxes is *

2. The mean amount of money is £*

3. The mean number of marbles in a glass is *

4.

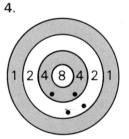 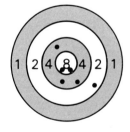 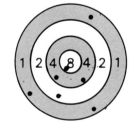

Pat's target   Reza's target   Lynne's target

a. What is Pat's total score?   b. What is Pat's mean score?
c. What is Reza's total score?   d. What is Reza's mean score?
e. What is Lynne's total score?   f. What is Lynne's mean score?

The mode

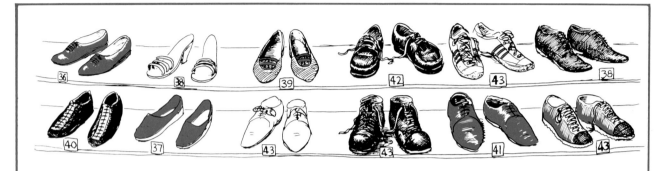

Shoe size 43 is the most popular in this display so we say that size 43 is the mode or modal value. We can use the mode or modal value as a type of average.

Exercise 11

1. What is the modal collar size?

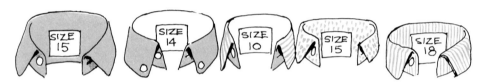

2. Write down the modal shoe size for these shoes.

3. Write down the mode for each of the three charts.

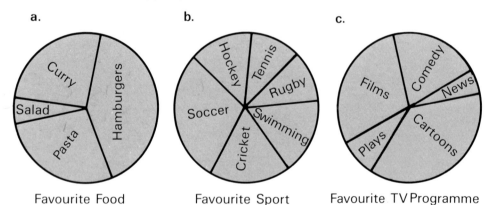

a. Favourite Food   b. Favourite Sport   c. Favourite TV Programme

4. Which of the four possible answers do you think is correct?

   a. The modal number of children in a family

       7   2   1   15

   b. The modal shoe size for women

       30   44   50   38

The median

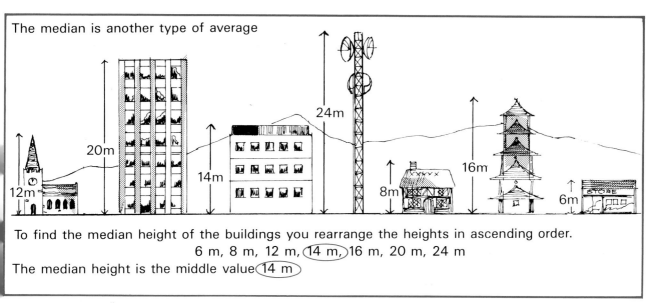

The median is another type of average

To find the median height of the buildings you rearrange the heights in ascending order.

6 m, 8 m, 12 m, 14 m, 16 m, 20 m, 24 m

The median height is the middle value 14 m

Exercise 12

1. Measure the lines below and find the median length

   a. ————————   b. ————————————————

   c. ——————————————————————   d. ——————————

   e. ——————————————————————   f. ————————————

   g. ——————————————————————

2. Sue is 120 cm tall, John is 115 cm tall, James is 97 cm tall
   Jason is 121 cm tall, Gene is 130 cm tall, Sara is 122 cm tall
   Phillipa is 150 cm tall, Paula is 117 cm tall and Corey is 129 cm tall.
   a. What is the median height of these children?
   b. Who has the median height?
   c. Does the median value give you a good idea of the typical height
   of these children?

3.

   a. What is the median shoe size?
   b. You are a shoe shop keeper. You want to stock a lot of shoes of
   the most typical shoe size. Would you find the median shoe size of
   your customers or the mode?

4. The numbers below are scores out of ten in a small maths test
   5, 5, 10, 1, 2, 3, 1, 7, 6, 6, 4, 10, 9, 6, 9, 8, 1

   a. How many pupils took the test?
   b. What was the median result?
   c. Would this be a fair mark to use as a pass mark? Why?

Exercise 13   Answer the questions below. Use the drawing below to help you.

1. **a.** What is the height of the tallest?
   b. What is the height of the shortest?
   c. What is the modal height?
   d. What is the median height?
   e. What is the mean height?
   f. Which of the three averages gives you the best idea of the typical height of the children?

Answer these questions.

2. Liza saved £24 in eight weeks.
   a. What was the average amount saved each week?
   b. Which type of average did you use to answer the question?

3. This is a list of wages paid by different shop-owners to children who do paper rounds.
   £2, £6, £8, £4, £7, £2, £5, £9
   a. What is the modal wage?
   b. Does the mode give a true idea of the average wage?
   c. Which type of average would you use to get an idea of the typical wage?

4. Eleven pupils take a maths test. The test was marked out of ten and the results are listed below:
   1, 7, 9, 2, 10, 9, 2, 1, 9, 10, 6
   a. What is the modal mark?
   b. What is the mean mark?
   c. What is the median mark?
   d. Which of the three averages tells you the most about the pupils' ability?

5. A group of children earn a mean wage of £9 per week in their weekend jobs. In total they earn £90, how many children are there?

6. If you were a dress maker and you wanted to know the average size of dress to make would you choose the median, mode or mean size of your customers as a guide?
   Remember your dresses must fit your customers.

Street maths 2: On the road

Ray Raver and the Sweet Tones released their new record, 'Keep Cool'. It sold 13,127 copies during the first week, and entered the Top Ten.

The Top Ten

Group	Title	Sales
The A.K. Band	Rock-et	16,792
T.C. Barret	Jean	11,052
Rupert Groove	Echo, Echo	14,859
The Crew	She's 18	10,237
Crystal Cleer	Wanting You	21,133
The Sweet Tones	Keep Cool	13,127
The Kidz	Warm Nights	13,688
Jail Breakers	Running Back	9,203
The Bio Band	City Sights	11,739
Mitch and Mo	Going West	15,572

Exercise 1

1. Round-off the sales figures for each band to the nearest 1000.

2. Rewrite the Top Ten in order of their sales figures. Put the band with the highest sales figures first.

3. How many more records did Crystal Cleer sell than the Sweet Tones?

4. How many records did the top three groups sell altogether?

5. How many groups sold more than 12,000 records?

The Sweet Tones' fan club sends a newsletter to all their fans.
The newsletter contains a profile of each group member.

THE SWEET TONES — PROFILE

Ray Raver	Bunny Grey	Leroy LeGrand	Eric Pluto
Age 33 years	Age 19 years	Age 18 years	Age 27 years
Height 1·83 m	Height 1·56 m	Height 1·91 m	Height 1·66 m
Weight 85 kg	Weight 56 kg	Weight 70 kg	Weight 74 kg
Lead singer	Keyboards—Saxophone	Drums	Guitar — Trumpet

Exercise 2

Using the profile above and the clues below answer the question 'who am I?'

1. I am shorter than Eric. I am ____
2. I weigh more than Leroy and less than Ray. I am ____
3. I am the tallest member of the group. I am ____
4. I am heavier than Eric. I am ____
5. I am the third tallest in the group. I am ____
6. I am older than 20, but I don't sing. I am ____
7. I am shorter than 1·70 m and I don't play the guitar. I am ____
8. I weigh more than 72 kg and I don't play the trumpet. I am ____

Exercise 3   Age

1. Ray is ____ years older than Leroy.

2. Eric is ____ years older than Bunny.

3. ____ is 14 years younger than Ray.

4. ____ is 9 years older than Leroy.

5. Eric is ____ years older than Leroy.

6. The age difference between the oldest and the youngest is ____ years.

Exercise 4   Weight

1. Ray is ____ kg heavier than Eric

2. Bunny is ____ kg lighter than Ray

3. Leroy is 15 kg lighter than ____

4. There is 14 kg difference between ____ and ____

Exercise 5   Height

1. Leroy is ____ cm taller than Eric.

2. Bunny is ____ cm shorter than Eric.

3. Ray is ____ cm taller than Eric.

4. The height difference beween the shortest and the tallest is ____ cm.

The tour

Ray Raver and the Sweet Tones begin their first tour of the U.S.A.
They arrive in New York and the first thing they do is
exchange their money for United States dollars ($)
They go to a bank and discover that they will get one dollar and 40
cents ($1·40) for each pound.

> For £1 you would receive $1·40
> For £10 you would receive $14
> For £100 you would receive $140

Exercise 6

Each member of the group exchanges some of their English money for
dollars. Copy and complete these sentences.

1. Ray cashes £100, he gets $ * * *   2. Leroy cashes £300, he gets $ * * *
3. Bunny cashes £380, she gets $ * * * 4. Eric cashes £261, he gets $ * * *

Exercise 7

The group go to buy food at Jumbo's Burger Palace. How much do they
each spend?

1. Ray's snack

= $3·00

= $0·60

= $0·60

Total = $ *

2. Leroy's snack

= $0·60

= $1·20

= $1·00

Total = $ *

3. Eric's snack

= $0·80

= $1·30

Total = $ *

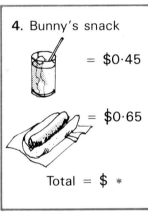

4. Bunny's snack

= $0·45

= $0·65

Total = $ *

Exercise 8

1. The four members of the group went for a bus ride.
   The fare was $1·25 each. What was the total cost?

2. The Sweet Tones went to the cinema. The total cost
   was $16·80. How much did they each pay?

3. Leroy bought a hamburger for $1·20. He paid for
   it with a $5 note. How much change did he get?

4. Bunny bought three magazines. They cost $1·20,
   $1·60 and $0·80. What was the total cost?

5. Eric bought two records at $3·85 each. How much was the total
   cost?

6. Ray went out to dinner. The bill came to $15·50.
   How much change did he get if he paid with $20?

7. The four members of the group each bought a hat.
   The hats were $3·35 each. What was the total cost?

Eric decided that it would be useful to compare the price of things.
He therefore drew up a conversion graph to translate dollars into pounds.

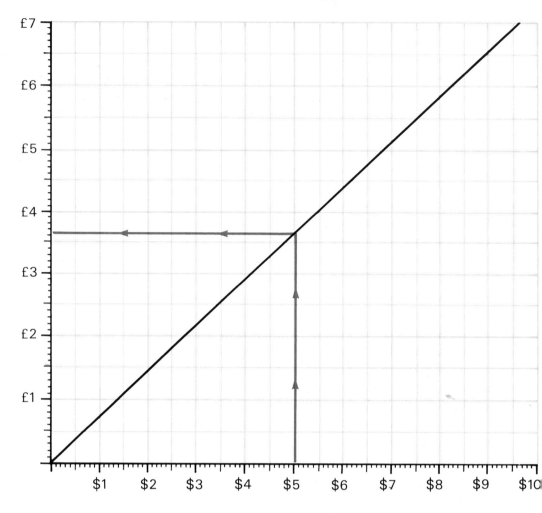

Using this conversion graph we see that $5 is about £3·60.

Exercise 9

Use the conversion graph to find out approximately how much these items cost in pounds and pence.

1. $2·00
2. $4·00
3. $5·00
4. $2·50
5. $9·00
6. $7·00
7. $4·50
8. $6·00
9. $3·50
10. $9·50

Exercise 10

Convert these amounts from pounds and pence to dollars and cents.

1. £2·00 is about $ *
2. £5·00 is about $ *
3. £6·00 is about $ *
4. £1·50 is about $ *
5. £4·50 is about $ *
6. £6·50 is about $ *

Scale

Ray Raver and the Sweet Tones are staying in Midchester at the start of their tour.

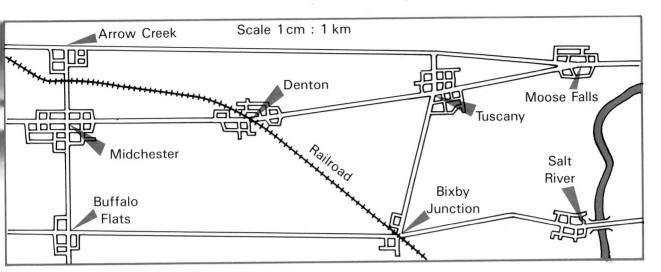

Scale 1 cm : 1 km

Arrow Creek
Denton
Moose Falls
Tuscany
Midchester
Railroad
Salt River
Buffalo Flats
Bixby Junction

Exercise 11

Before the concert Ray Raver decided to go for a drive to Denton.
He looked at the map and said, 'It doesn't show how far Denton is from Midchester'.
'Use the scale'. Leroy told him. 'What?' said Ray.

Leroy explains.
'If you measure the distance between Midchester and Denton on the map, you find they are 5 cm apart. The scale tells us that each centimetre on the map represents 1 kilometre on the road'. Midchester is 5 cm from Denton on the map, so Ray will have to drive 5 km to get from Midchester to Denton.

Use a ruler to measure the distance between the blue markers. Convert your answers into kilometres. Remember 1 cm : 1 km

1. Midchester is * km from Buffalo Flats.

2. Midchester is * km from Arrow Creek.

3. Tuscany is * km from Moose Falls.

4. Tuscany is * km from Bixby Junction.

Exercise 12

1. How far is it from Moose Falls to Arrow Creek by the shortest route?

2. How far is it from Buffalo Flats to Salt River by the shortest route?

3. How far is it from Midchester to Moose Falls by the shortest route?

4. If Ray drives from Midchester to Buffalo Flats, Bixby Junction, Tuscany and then back to Midchester, how many kilometres has he travelled?

5. How far is it from Salt River to Moose Falls by the shortest route?

The Sweet Tones leave Midchester on the next stage of their tour. They go to Austin City.

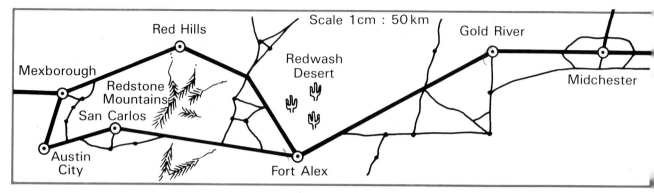

On this map 1 cm represents 50 km. The scale is 1 cm : 50 km.

Exercise 13

The Sweet Tones travel from Midchester to Austin City in their tour coach.

1. The journey from Midchester to Gold River is * km.
2. The journey from Gold River to Fort Alex is * km.
3. The journey from Fort Alex to San Carlos is * km.
4. The journey from San Carlos to Austin City is * km.
5. The total journey is * km.

Exercise 14

Using the scale above 0·5 cm represents 25 km.

1. How far is it from Austin City to Mexborough by the shortest route?

2. How far is it from Mexborough to Red Hills by the shortest route?

3. How far is it from Red Hills to Fort Alex by the shortest route?

4. How far is it from Gold River to Red Hills, by the shortest route?

5. There is a direct air service between Red Hills and Gold River. What distance does the aeroplane travel on this journey?

Exercise 15

Redraw this map to scale. Use a scale of 1 cm : 10 km.

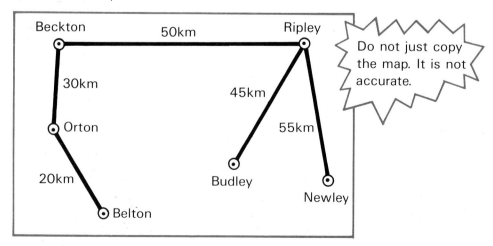

It is the last concert of the Sweet Tones' tour.
5,766 people watch their final performance in the U.S.A.

Exercise 16

ATTENDANCES AT CONCERTS	
May 6th	3,424
May 7th	5,834
May 8th	4,616
May 9th	4,209
May 11th	3,974
May 12th	5,282
May 13th	5,575
May 14th	3,121
May 15th	5,065
May 16th	5,766

1. On which date did most fans attend a Sweet Tones' concert?

2. On which date did least fans attend a Sweet Tones' concert?

3. Re-write the attendance figures in order, starting with the largest number.

Exercise 17

1. The cost of a ticket for each concert was $5. How many dollars were received when 3121 fans came to see the group?

2. The cost of touring was £4,087 for each member of the group. What was the cost for the four of them on tour?

3. The group earned $12,812. How much did each member get?

4. The group travelled 18,808 km by air, 2,698 km by coach, 197 km by train and 64 km by car. How many kilometres did they travel in total?

5. Bunny signed 1,627 autographs, Ray signed 27, Leroy signed 942 and Eric signed 784. How many autographs were signed in total?

6. If 14 songs were sung in 84 minutes, how long, on average did each song last?

7. In the middle of their 84 minute act, the group take a 15 minute break. How many hours and minutes does the act last for?

8. During the tour, fans could buy Sweet Tone tee-shirts. 163 blue ones were sold, 198 red shirts, 96 green shirts, and 219 yellow shirts. How many were sold in all?

Section 17   **Coordinates**

The numbered lines are called axes.
The x-axis runs across the bottom of the grid.
The y-axis runs up the side of the grid.

Remember that the first number in a pair of coordinates is on the x-axis.
The second number is on the y-axis.

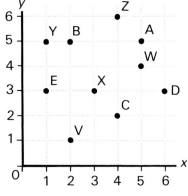

Exercise 1

What letters will you find at these coordinates?
1. (4, 2)   **2.** (2, 5)   **3.** (5, 5)   **4.** (1, 3)   **5.** (6, 3)

Give the coordinates for these points.
6. V   **7.** W   **8.** X   **9.** Y   **10.** Z

Exercise 2

Give the coordinates for the four corners of this rectangle.

     A (∗, ∗)
     B (∗, ∗)
     C (∗, ∗)
     D (∗, ∗)

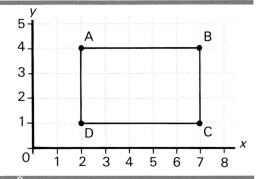

Exercise 3

1. This grid is incomplete.
Can you give the coordinates for corner B?

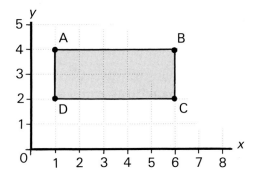

2. This grid is incomplete.
Can you give the coordinates for corner B?

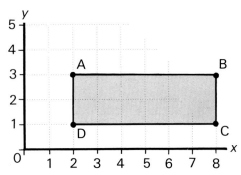

Exercise 4

1. This grid is incomplete.
   Write down the coordinates
   for points A, B and C

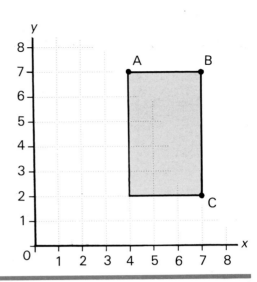

Exercise 5

The rectangles on this grid
are identical. Each rectangle
is four units long and
three units high.

What are the coordinates
of point Z.

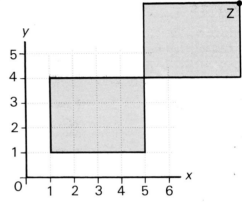

Exercise 6

The rectangles on this grid
are identical.

What are the coordinates
of points R, S, T.

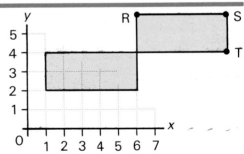

Exercise 7

1.

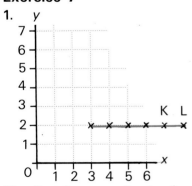

The first four coordinates for this
line are (3, 2) (4, 2) (5, 2) (6, 2)
What are the coordinates of K and L?

2.

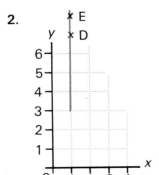

Give the coordinates
of points D and E.

3.

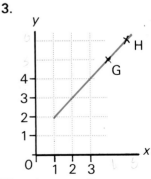

Give the coordinates
of points G and H.

In this diagram the y-axis has been extended down below the x-axis. Numbers used in positions below the x-axis have a minus sign in front of them.

> The coordinates of point A are (2, −4)
> 2 along the x-axis.
> −4 on the y-axis.

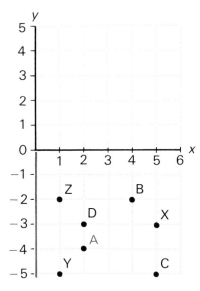

Exercise 8

What letter will you find at these coordinates?

1. (4, −2)   4 along the x-axis
            −2 on the y-axis

2. (2, −3)   2 along the x-axis
            −3 on the y-axis

3. (5, −5)   5 along the x-axis
            −5 on the y-axis

Give the coordinates for these points

4. X   **5.** Y   **6.** Z

In this diagram the x-axis has been extended to the left of the y-axis.

Point A is at coordinates (2, 3)
        2 on the x-axis
        3 on the y-axis

Point B is at coordinates (5, −2)
        5 on the x-axis
        −2 on the y-axis

Point C is at coordinates (−3, 4)
        −3 on the x-axis
        4 on the y-axis

Point D is at coordinates (−4, −3)
        −4 on the x-axis
        −3 on the y-axis

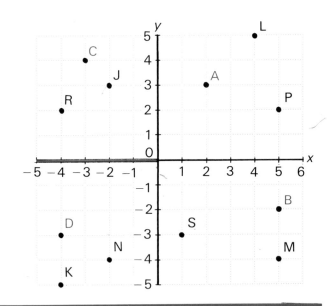

Exercise 9

What letters will you find at these coordinates?

1. (5, 2)   **2.** (4, 5)   **3.** (1, −3)   **4.** (5, −4)

5. (−2, 3)   **6.** (−4, 2)   **7.** (−2, −4)   **8.** (−4, −5)

The rectangle has a corner in each of the four quarters of the graph.

Corner A has coordinates (3, 4)
Corner B has coordinates (−3, 4)
Corner C has coordinates (−3, −4)
Corner D has coordinates (3, −4)

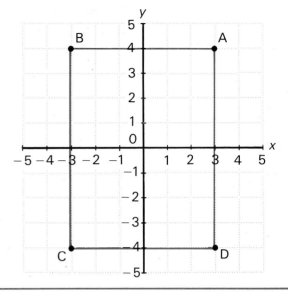

Exercise 10

1. Draw five grids like the one above

2. Plot these five sets of points, one onto each of the five grids, joining them up as you plot them

    set **a**   (5, 3) (−5, 1) (2, −4) (5, 3)

    set **b**   (4, 0) (0, 4) (−4, 0) (0, −4) (4, 0)

    set **c**   (3, 2) (2, 3) (−4, −2) (−3, −3) (3, 2)

    set **d**   (3, 2) (−3, 2) (−3, −2) (3, −2) (3, 2)

    set **e**   (−2, 5) (−5, 3) (−2, −4) (1, 3) (−2, 5)

3. What five shapes have you drawn?

Exercise 11

By finding the position of letters given by the coordinates you should be able to spell out a tongue twister.

(1, 4) (−1, 1) (−2, 3)

(4, 1) (2, −3) (−5, −4)

(−3, −2) (2, 3) (4, −1) (−4, 4)

(−3, −2) (4, 4) (4, −4) (−4, 2) (−2, 3) (1, −1)

(4, 1) (−5, −1) (5, −3) (−4, 2)

(4, 1) (−5, 5) (2, 3) (−4, 2) (−2, 3)

(−3, −2) (4, −1) (4, −4) (1, 2) (−2, −4)

(−2, 5) (1, −3) (4, −4) (−4, 2) (−2, 3)

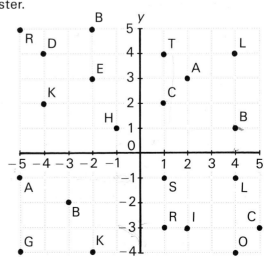

Reflection

Look at the grid. The grey rectangle is a mirror image of the white rectangle.

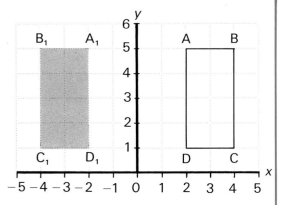

The y-axis acts as a mirror.

Corner A is reflected to corner A_1 (2, 5) → (−2, 5)

Corner B is reflected to corner B_1 (4, 5) → (−4, 5)

Corner C is reflected to corner C_1 (4, 1) → (−4, 1)

Corner D is reflected to corner D_1 (2, 1) → (−2, 1)

Exercise 12

1. Draw the grid on centimetre squared paper.

2. Draw the rectangle on the graph and draw its reflection in the y-axis.

3. List the coordinates of the corners of the reflection.

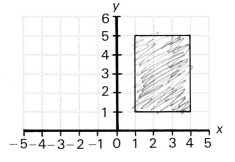

Exercise 13

1. Draw a grid like this on centimetre squared paper.

2. Plot these points on the grid and join them up to form a rectangle (0, 2) (1, 1) (5, 5) (4, 6)

3. Use the y-axis as a mirror.

4. List the coordinates of the corners of the reflected rectangle.

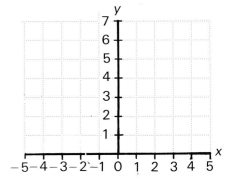

Exercise 14

1. Draw a grid like the one above.

2. Plot these points onto the grid: (−5, 1) (−5, 6) (−1, 2)

3. Join up the points to form a triangle.

4. Use the y-axis as a mirror, and list the coordinates of the corners of the reflected triangle.

The grey triangle is the mirror image of the white triangle. The reflection uses the x-axis as the mirror.

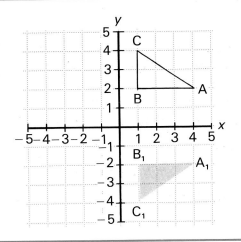

Point A is reflected to point A_1 (4, 2) → (4, −2)
Point B is reflected to point B_1 (1, 2) → (1, −2)
Point C is reflected to point C_1 (1, 4) → (1, −4)

Exercise 15

1. Draw a grid like this one on centimetre squared paper.

2. Draw the shape on the grid.

3. Use the x-axis as a mirror.

4. List the coordinates of A_1, B_1, C_1, D_1 of the reflected shape.

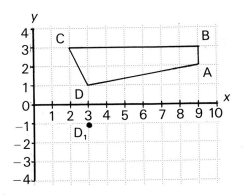

Exercise 16

1. Draw this grid on centimetre squared paper.

2. Draw the shape on the grid.

3. Use the x-axis as a mirror.

4. List all the coordinates of the reflected points A_1, B_1, C_1, D_1, E_1, F_1, G_1, H_1.

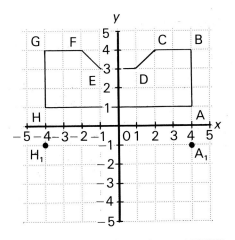

Exercise 17

1. Draw a grid like the one above and plot these points on the grid.
   (−5, 0) (−3, 4) (0, 1). Join them up to make a triangle.

2. Reflect the triangle in the x-axis and write down the coordinates of the reflected points.

Exercise 1

Graph A shows the average monthly temperature

1. Which was the hottest month?
2. Which was the coolest month?
3. What was the average temperature in May?

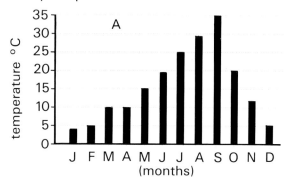

Exercise 2

Graph B shows the average monthly rainfall

1. Which were the driest months?
2. Which was the wettest month?
3. In which month did 7 cm of rain fall?

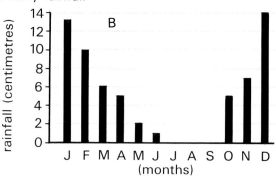

Exercise 3

Graph C shows the average number of tourists

1. In which month did most tourists visit?
2. In which months did least tourists visit?
3. In which month were there 800 visitors?

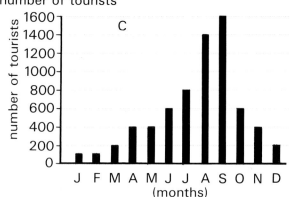

Exercise 4

Use all the graphs to answer these questions.
1. In which month did 400 tourists visit, and 2 cm of rain fall?
2. Why do you think so many tourists left between September and October?
3. How many visitors came during the three coolest months?

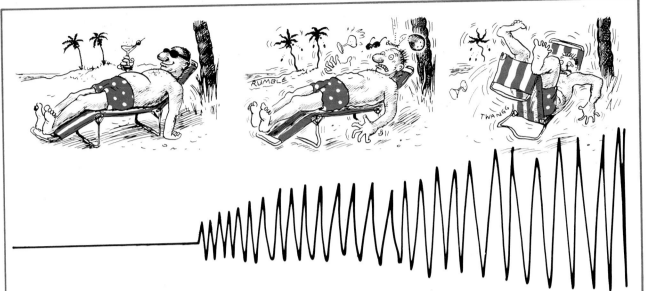

The Seashell Islands suffer from earth tremors. These tremors are measured on a SEISMOGRAPH. When the ground shakes it rocks a pen on a moving sheet of paper. The more violent the tremor the more violently the pen rocks.

Here is a seismograph record for 8 days.

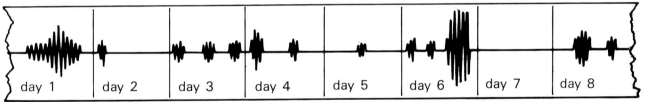

day 1   day 2   day 3   day 4   day 5   day 6   day 7   day 8

Exercise 5

1. How many tremors were there on day 4?

2. How many tremors were there on day 3?

3. On which day were there no tremors?

4. Which day had the tremor lasting the longest time?

5. Which day had the most violent tremors?

Exercise 6

Here is a graph of earth tremor activity on day 1.

Using the seismograph record above draw a graph for each of the other days.

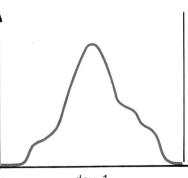

great activity

no activity

day 1.

stage 1   stage 2   stage 3

This graph shows how brightly the firework burns from the moment it is lit.

Stage 1 It starts off dull becoming brighter.

Stage 2 It burns very brightly for a while.

Stage 3 It burns out and becomes dark again.

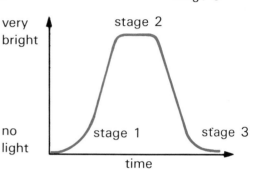

Exercise 7   Match the graphs with the correct statements below.

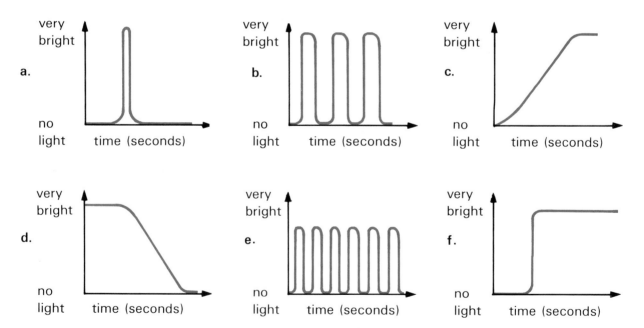

1. Car headlights flashing three times

2. A light is switched on quickly in a dark room

3. A flash light flashes once

4. Cinema lights being dimmed slowly

5. A car indicator flashing many times

6. Cinema lights being brightened slowly

Exercise 8

Match the sentences below with the correct temperature graph.
For example, sentence **a.** matches with graph **1**.

a. The temperature starts from 0°C and climbs steadily.

b. The temperature falls steadily from 20°C to 0°C.

c. The temperature rises and falls very quickly.

d. The temperature rises, stays the same for a while and climbs again.

e. The temperature stays the same at 0°C, then rises.

f. The temperature stays the same throughout.

g. The temperature stays the same then falls steadily.

h. The temperature rises, stays warm and then falls steadily.

j. The temperature starts above 0°C and increases steadily.

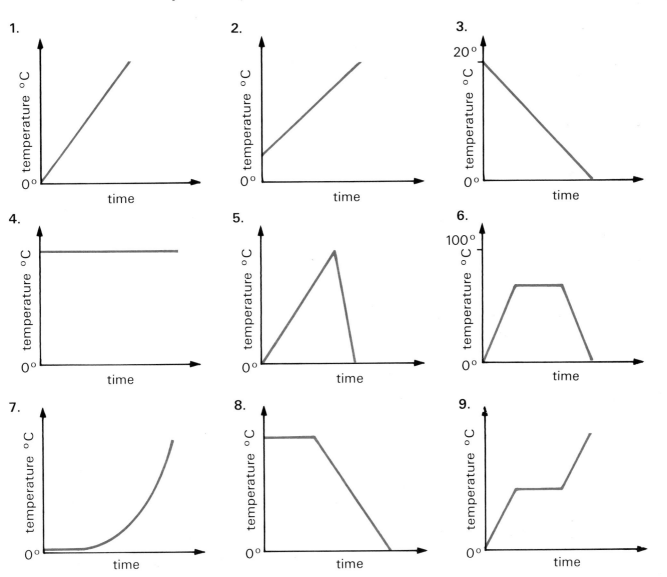

Activity graph

The graph below shows how active Linda was during a full day on holiday at Pontlins Holiday Camp.

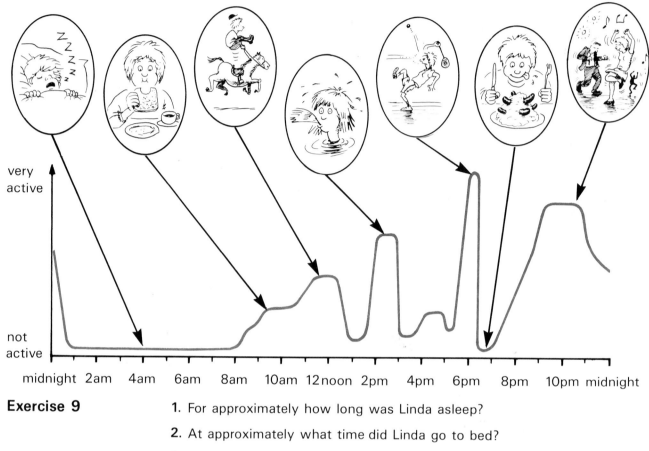

Exercise 9

1. For approximately how long was Linda asleep?

2. At approximately what time did Linda go to bed?

3. Was the build up of activity to horse-riding gradual or sudden?

4. What is Linda doing when she is most active?

5. Which activity, other than sleeping, did Linda spend the longest doing?

6. Between horse riding and swimming was Linda resting or busy?

Exercise 10

On axes like the ones below draw an activity graph for your own school day

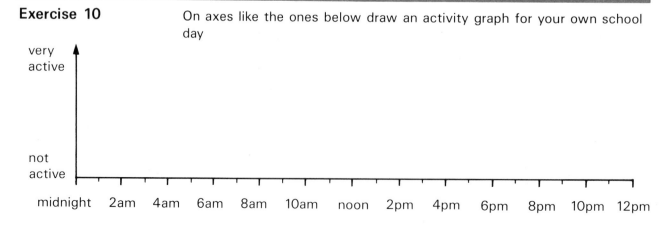

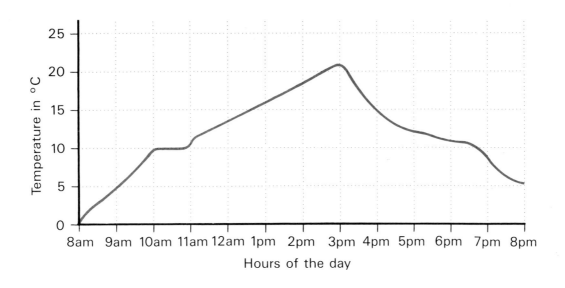

Exercise 11

Answer these questions about the graph above.

1. Is the temperature rising or falling between 8am and 3pm?

2. What is happening to the temperature between 3pm and 7pm?

3. What is happening to the temperature between 10am and 11am? It is
   a. rising
   b. staying constant
   c. falling?

4. At what times did the temperature reach 15°C?

5. When did the temperature reach its highest?

6. What was the highest temperature reached?

Exercise 12

Answer these questions about the graph below.

1. What was the temperature at 6am?

2. What was the highest temperature reached approximately?

3. What was the lowest temperature reached approximately?

4. About what time did the temperature reach its highest?

5. About what time did the temperature reach its lowest?

6. Did the temperature fall quicker than it rose?

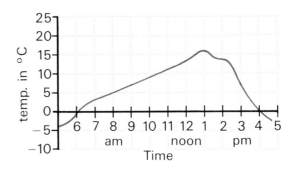

Alex has saved £40 to spend on clothes. He wants to buy a pair of training shoes, a pair of jeans and a jumper in the leisure wear sale. Alex decides that these are the items he likes:—

Exercise 1

How much cheaper in the sale are:—

1. The Ajax trainers?   2. The Classic jeans?   3. The DeVare jumper?
4. The Slimfit jeans?   5. The Pronto trainers?   6. The Keeval jumper?

Exercise 2

1. How much would the most expensive trainers, jeans and jumper cost together?

2. How much would the least expensive trainers, jeans and jumper cost together?

3. Choose a pair of jeans, a pair of trainers and a jumper that Alex could afford with his £40.

4. Alex bought the Ajax trainers, Fargo jeans and the Ricco jumper. How much did it cost him?

5. How much change was he left with?

6. If Alex had bought the least expensive trainers, jeans and jumper, how much change would he have?

Percentages

In each shop is a sign like this. **10%** The sign is telling us of a price reduction.
This sign % means per cent.
Per cent means per hundred.

10% means $\frac{10}{100}$ or $\frac{1}{10}$. A 10% reduction means reducing the price by $\frac{1}{10}$.

Exercise 3

These watches are being reduced by 10%. Give the amount of each reduction. The first one has been done for you.

1. 10% of £20 $= \frac{1}{10}$ of £20
$\qquad\qquad = £20 \div 10$
$\qquad\qquad = £2$

2.  £50 **3.**  £30 **4.**  £40 **5.**  £10 **6.**  £20

7. £70   **8.** £90   **9.** £100   **10.** £110   **11.** £60   **12.** £200

Exercise 4

A 20% reduction means reducing the price by $\frac{2}{10}$ or $\frac{1}{5}$.
If these amounts are all reduced by 20%, give the amount of each reduction.

1. 20% of £10 $= \frac{1}{5}$ of £10.
$\qquad\qquad = £10 \div 5$
$\qquad\qquad = £2$

2. £20   **3.** £30   **4.** £60   **5.** £40   **6.** £15   **7.** £25

Exercise 5

A 25% reduction means reducing the price by $\frac{25}{100}$ or $\frac{1}{4}$.
If these amounts are all reduced by 25%
a. find the reduction   **b.** what is the new price?

1. £40   **2.** £16   **3.** £20   **4.** £24   **5.** £80   **6.** £32

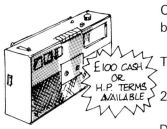

Carol wants to buy a cassette player. She can buy it with cash or buy it by paying fixed instalments over a period of time (hire purchase).

The cost of the machine is £100 cash

or

20% of the cost as a deposit, and £10 a month for 10 months.

Deposit   20% of £100   = £20
Instalments £10 x 10 months = £100
        Total cost   = £120

It is therefore £20 dearer to buy this machine on hire purchase.

Exercise 6

Carol wishes to buy this machine which costs £50 cash.
The hire purchase arrangements are
   10% deposit and £10 a month for 5 months.

1. Copy and complete these calculations to find the cost of payment by hire purchase

Deposit   10% of £50   = *
Instalments £10 x 5 months = *
        Total cost   = *

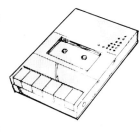

2. How much more does it cost to buy the cassette player on hire purchase?

Tommy wishes to buy this television costing £200 cash.
The hire purchase arrangements are
   20% deposit and £20 a month for 12 months.

3. Copy and complete these calculations to find the cost of payment by hire purchase

Deposit   20% of £200   = *
Instalments £20 x 12 months = *
        Total cost   = *

4. How much more does it cost to buy the television by hire purchase?

Mrs Lewis wishes to buy this freezer costing £150 cash.
The hire purchase arrangements are
   10% deposit and £15 a month for 12 months.

5. What is the total cost of buying the freezer by hire purchase?

6. How much more does it cost to buy the freezer by hire purchase?

Marco and Jean are shopping in the market.
They are buying fruit and vegetables.

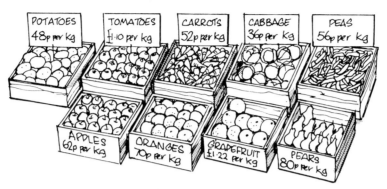

Exercise 7

How much will Marco and Jean pay for:

1. two kilograms of cabbage?
2. two kilograms of tomatoes?
3. two kilograms of apples?
4. three kilograms of potatoes?
5. three kilograms of oranges?
6. three kilograms of carrots?

There are 1000 grams in a kilogram.

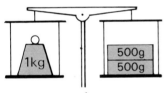

500g is $\frac{1}{2}$ of a kg
$\frac{1}{2}$ kg + $\frac{1}{2}$ kg = 1 kg

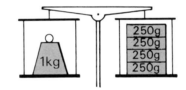

250g is $\frac{1}{4}$ of a kg
$\frac{1}{4}$ kg + $\frac{1}{4}$ kg + $\frac{1}{4}$ kg + $\frac{1}{4}$ kg = 1 kg

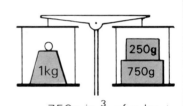

750g is $\frac{3}{4}$ of a kg
$\frac{3}{4}$ kg + $\frac{1}{4}$ kg = 1 kg

Exercise 8

How much will Marco and Jean pay for:

1. 500 g of oranges?
2. 500 g of tomatoes?
3. 500 g of pears?
4. 500 g of apples?
5. 500 g of peas?
6. 500 g of carrots?

Exercise 9

How much will Marco and Jean pay for:

1. 250 g of potatoes?
2. 250 g of cabbage?
3. 250 g of peas?
4. 750 g of pears?
5. 750 g of potatoes?
6. 750 g of cabbage?

Exercise 10

What weights are being used to balance these scales?

1.

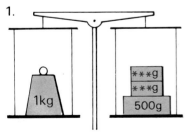

2.

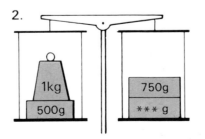

3.

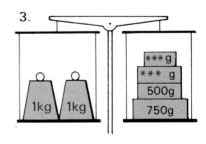

Exercise 11

Every year at Christmas, Mr. Farr sends sweets to the children's ward of the local hospital.

1. How many kg of sweets did he send to the hospital if he packaged up:

> 450g of nut crunch
> 800g of chocolate fudge
> 600g of bitter fruit drops
> 750g of assorted toffees
> 350g of cola cubes
> 500g of chocolate buttons

2. Mr. Farr sends some sweets to the local junior school. How many kilograms did he send if he packed up:

> 250g of chocolate eclairs
> 550g of strawberry chews
> 300g of bitter lemons
> 200g of chocolate limes
> 450g of liquorice allsorts

3. Mr. Farr has to re-order stock. Here is his order form.

How many kilograms of sweets is he ordering?

> **ORDER FORM**
> FROM: Farr's Sweetshop TO: Sweet Supplies
>
> Toffee Cubes 1.50kg ColaCubes 1.50kg
> Chocolate Limes 1.75kg ButterLemons 1.25kg
> Nut Crunch 2.25kg NutToffees 1.50kg
> Choc-Mints 1.00kg Fizzy Chews 2.50kg

Exercise 12

What weight will be shown on these scales?

1.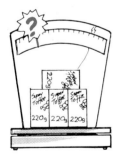

2.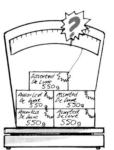

3.

4.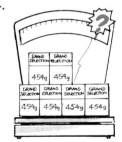

How many grams does each pack weigh?

5.

6.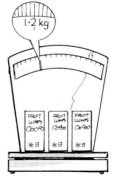

7.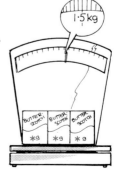

8.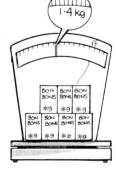

Capacity

Capacity is a measure of volume. When you buy liquids like petrol wine, or lemonade the amount that you buy is measured in litres (*l*).

A litre is just a little less liquid than 2 pints. For measuring small amounts of liquid we can use millilitres (m*l*) There are 1000 millilitres in a litre.

1 litre = 1000 ml,
$\frac{1}{2}l$ = 0·5 *l* = 500 ml
$\frac{1}{4}l$ = 0·25 *l* = 250 ml
$\frac{3}{4}l$ = 0·75 *l* = 750 ml

A can of soft drink is usually 0·44 *l* or 440 ml

A bottle of wine holds about 0·7*l* or 700 ml

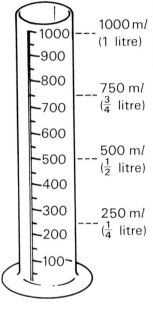

1000 ml (1 litre)

750 ml ($\frac{3}{4}$ litre)

500 ml ($\frac{1}{2}$ litre)

250 ml ($\frac{1}{4}$ litre)

Exercise 13

Use these drawings to answer the questions below.
Say whether these statements are true or false.

1. 1 bottle of cough mixture is less than 1*l*.

2. 1 bottle of ink is more than 250 m*l*.

3. There is more liquid in the Orange Crush bottle than there is in the bottle of cough mixture.

4. One bottle of Cream Soda contains the same amount of liquid as two cartons of Choc Milk.

5. A can of Fizz and a can of Shandy together contain more liquid than a bottle of Cream Soda.

6. The capacity of two cans of Fizz is the same as one bottle of cough mixture.

7. It would take exactly twenty bottles of ink to fill 1 bottle of Cream Soda.

8. The contents of a can of Shandy will fit into an empty Choc Milk carton.

Robin, Judy and Kevin met in the street. Each had bought a bottle of cola because it was a hot day.

Who has best value for money? To answer this question we must calculate how much each person paid for a common amount of coke. In this case the common amount is 0·5 or $\frac{1}{2}$ l of cola.

Robin paid 64p for 1l.   0·5l cost him 32p.
Judy paid 84p for 1·5l.   0·5l cost her 28p.
Kevin paid £1·20 for 2l.   0·5l cost him 30p.

Judy got the best value for money, because she got 0·5l for the least amount of money.

Exercise 14

Which offers best value?

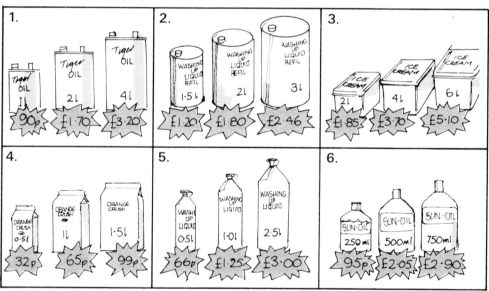

Exercise 15

Use the pictures above to help you answer these questions.

1. In picture 2, how many more mls is the medium size of liquid than the small size?

2. In picture 4, how many mls are there in the large carton of orange?

3. In picture 6, how many small bottles of Sun Oil have the same capacity as the large bottle?

A pie chart is a type of circular graph. It is a way of showing information.

From this graph it is easy to see that cola is the most popular drink because cola takes up the biggest single piece of the chart.

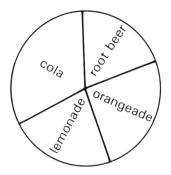

Pie chart showing popular drinks

Exercise 1

Look at the pie chart and estimate your answers

1. How many ways of using pocket money are shown on the chart?

2. Which is the most popular way of spending money?

3. Which is the least popular way of spending money?

4. Do pupils spend more money on concerts than going to the cinema?

5. Do pupils save more money than they spend on sweets and drinks?

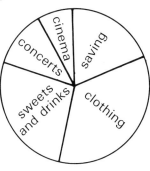

Pie chart showing how pupils use their pocket money

Exercise 2

1. How many different types of programmes are shown on the chart?

2. Make a list of the different types of programmes shown. Put the list in order from the most popular to the least popular.

3. Which two types of programme are equally popular?

4. Which is the least popular type of programme?

5. Re-draw the pie chart in your book showing how you think it would look if you did the survey in your own class.

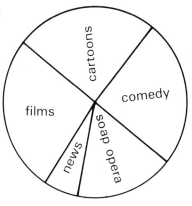

A pie chart showing popular television programmes

This pie chart has been divided into ten equal portions.
We can say that each portion is 10% of the whole chart.

From the chart we know that 10% of children cycle to school, 20% travel by train, 30% travel by bus and 40% walk.

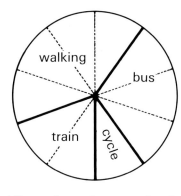

Ways of coming to school

Exercise 3

Answer these questions accurately

1. What percentage of people prefer Britain for their holiday?

2. Which other two holiday places are preferred by 10% of the people?

3. What percentage of people prefer holidays in the U.S.A.?

4. Is it true to say that 50% of people prefer either the U.S.A. or Greece?

5. What percentage of people choose either Spain or Greece?

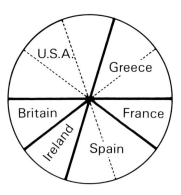

Popular holiday places

Exercise 4

Answer these questions accurately

1. Which is the most popular pet?

2. What percentage of pet owners keep this animal?

3. What percentage of pet owners keep rabbits?

4. Is it true to say that 50% of pet owners keep either dogs or rabbits?

5. How many times more people keep dogs than rats?

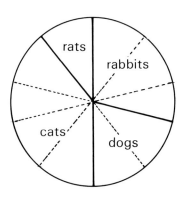

Popular pets

The pie chart has been divided into ten equal parts. Each part is $\frac{1}{10}$ or 10% of the whole chart.

50 people were questioned about their favourite training shoes.

10% of the people prefer Trako trainers

$$10\% \text{ or } \tfrac{1}{10} \text{ of } 50 = 50 \div 10$$
$$= 5 \text{ people}$$

20% of people prefer Funlop trainers

$$20\% \text{ or } \tfrac{2}{10} \text{ of } 50 = (50 \div 10) \times 2$$
$$= 10 \text{ people}$$

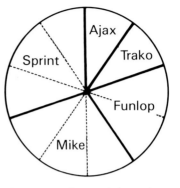

Most popular training shoes

Exercise 5

In the survey of most popular comic papers, 80 people were questioned.

1. How many people are represented by 10% of the chart?

2. How many people prefer 'My Girl' magazine?

3. How many people prefer the 'Charger'?

4. How many more people prefer the 'Whizz Kid' comic to the 'Deeno'?

5. How many like either the Deeno or the Charger?

Most popular comic papers

Exercise 6

150 people were questioned in this survey.

1. How many votes are represented by 10% of the chart?

2. How many people voted for
   a. Colin   b. Roy

3. How may people voted for Rita?

4. How many people voted for
   a. Maria   b. Carol

5. Who would have been elected?

6. How many more votes did the three girls receive than the boys?

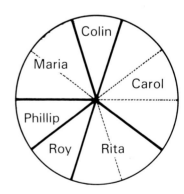

Votes for the School Council representative

Review 3

A. Symmetry

1. Look at these painted eggs. Which of them are drawn symmetrically?

a.    b.    c.    d.    e.

2. Copy these shapes. Put the lines of symmetry onto them.

a.   3cm   5cm

b. 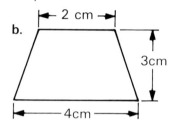   2 cm   3cm   4cm

B. Problems

1.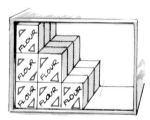

    a. How many bags of flour are in this box?

    b. How many bags would be in a full box?

2.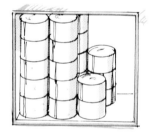

    a. How many tins are in this box?

    b. How many tins would be in a full box?

Complete these problems so that they are correct.

3.

a.
```
  23 *
+ 1 * 2
------
  386
```

b.
```
  3 * 5
+ 34 *
------
  679
```

c.
```
  458
+ 2 * *
------
  682
```

d.
```
  * 6 *
+ 4 * 7
------
  693
```

e.
```
  685
- 3 * *
------
  361
```

f.
```
  5 * 7
- 33 *
------
  254
```

g.
```
  6 * 9
- * 53
------
  42 *
```

h.
```
  760
- 3 * *
------
  444
```

i.
```
  23 x
    *
------
  46
```

j.
```
  3 * x
    2
------
  68
```

k.
```
  2 * x
    3
------
  69
```

l.
```
  2 * * x
      3
------
  696
```

m.
```
    23
 * )69
```

n.
```
    43
 2)8 *
```

o.
```
    24
 2) * *
```

p.
```
   * 04
 3)61 *
```

C. Angles in a circle

What would the tank be aiming at if the barrel pointed at these bearings? (Use a ruler to help you.)

1. 060°
2. 130°
3. 250°
4. 010°
5. 310°

What are the bearings for these targets?

6. The Radio Unit
7. Target No. 3.
8. Target No. 2.
9. The Old House.
10. The Helicopter.

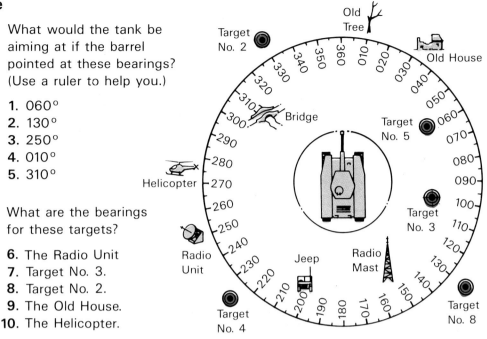

11. How many degrees of turn are there between the Old Tree and Target No. 5?

12. How many degrees of turn are there between the Old House and Target No. 3?

13. How many degrees of turn are there between Target No. 8 and the Jeep?

14. How many degrees of turn are there between the Helicopter and Target No. 2?

15. How many degrees of turn are there between the Radio Unit and the Old House?

D. Averages

| median The median is the middle value in an array. | mode The mode is the most common value in an array. | mean The mean is the total of all the values divided by the number of values. |

1. Find the median value in this group of test results
   60 43 47 61 35 53 49 32 59

2. Find the mode of this array of shoe sizes
   43 41 39 39 41 37 39 35 45

3. Find the mean for this array of scores
   6, 7, 6, 8, 8

4. Find the mean for these amounts of money
   20p 10p 12p 8p 15p 1p

E. Coordinates

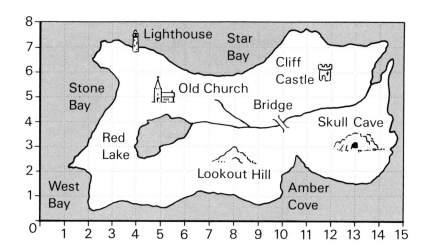

Re-write this passage replacing two coordinates with place names from the map.

They landed at (2, 2) and went to (4, 3) where they stopped to drink. From (4, 3) they made their way to (8, 3). From the summit they could see (12, 6) and the (4, 7). They left (8, 3) and walked to (13, 3) where they had lunch. On the way back to (2, 2) they crossed a (10, 4) and passed by an (5, 5)

F. Solids

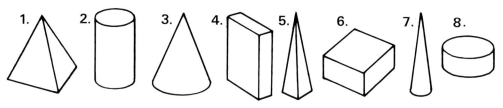

These solids have been redrawn below. The views have been changed. Copy the table and   **a.** match the solids   **b.** name them. The first has been done for you.

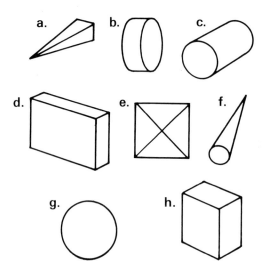

matching pair	Name
1e	Pyramid

G. Percentages

A 30% reduction means reducing the price by $\frac{3}{10}$.
If these amounts are all reduced by 30%, give the amount of each reduction.

1. £10   **2.** 50p   **3.** £2·50   **4.** 20p

5. £700   **6.** £60   **7.** £40   **8.** £1·50

9. Answer the following questions
   a. what is 10% of £40?   **b.** what is 50% of £200?
   c. what is 25% of £50?   **d.** what is 20% of £90?

H. Reflections

1.a. Copy the grid and draw the triangle ABC onto the grid.

   b. Use the y-axis as a mirror, and draw the reflection of the triangle.

   c. Write down the coordinates of the reflection, A_1, B_1, C_1.

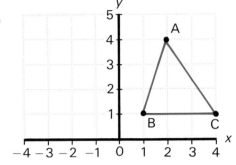

2.a. Copy the grid and draw the shape ABCD onto the grid.

   b. Use the x-axis as a mirror and draw the reflection. Write down the coordinates of the reflected shape (A_1, B_1, C_1, D_1)

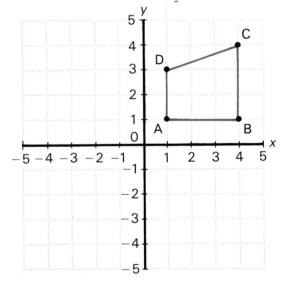

J. Decimals

Copy and find the answers to the following calculations.
(Estimate your answers before you do your calculations.)

   1. 5·2 kg + 9·7 kg − 3·8 kg = *   **2.** 8 kg − 7·2 kg = *

   3. 3·2 m + 4·9 m + 10 m = *   **4.** 0·2 m + 1·5 m + 1·3 m = *

   5. 3·5 m − 1·20 m = *   **6.** 9·1 m + 10·5 m − 13 m = *

   7. 4·2 x 4 = *   **8.** 3·1 x 3 = *   **9.** 5·2 x 4 = *   **10.** 6·2 x 3 = *

   11. 2·5 x 3 = *   **12.** 0·9 x 2 = *   **13.** 16·3 x 3 = *   **14.** 15·6 x 4 = *

   15. 4·8 ÷ 2 = *   **16.** 3·9 ÷ 3 = *   **17.** 15·3 ÷ 3 = *   **18.** 2·7 ÷ 9 = *

K. Probability

Answer these questions.

1. When tossing a coin, what is the probability of it being a 'head'?

2. What is the probability of getting a 'tail'?

For the questions below, use a dice or a picture of a dice to help you

3. What is the probability of rolling a 'four'?

4. What is the probability of rolling a 'six'?

5. What is the chance of rolling a 'two' or a 'one'?

6. What is the chance of rolling a 'three' or a 'six'?

7. What is the probability of rolling a 'one', a 'two' or a 'three'?

8. What is the probability of rolling a number bigger than 'four'?

L. Graphs

This graph shows the number of hours the sun shone on different days in a week

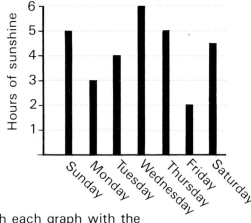

1. On which day did the sun shine longest?

2. On which day did the sun shine for 3 hours?

3. For how long did the sun shine on Sunday?

4. Below are three graphs. Match each graph with the statement that describes it.

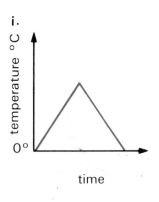
i.

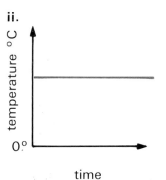
ii.

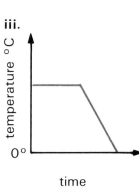
iii.

a. The temperature stays constant throughout

b. The temperature rises steadily then falls steadily

c. The temperature is constant, then falls steadily